# Avantgarde Reliability Implications in Civil Engineering

*Edited by Maguid Hassan*

Published in London, United Kingdom

Avantgarde Reliability Implications in Civil Engineering
http://dx.doi.org/10.5772/intechopen.102292
Edited by Maguid Hassan

Contributors
Shima Rahmani, Masoud Ebrahimi, Elyas Fadakar, Nicholas Eze, Ekene Ozioko, Johnpaul Nwigwe, Folahan Peter Peter Ibitoye, Dagan Bakun-Mazor, Karla Saldaña Ochoa, Maguid Hassan



First published in London, United Kingdom, 2023 by IntechOpen
IntechOpen is the global imprint of INTECHOPEN LIMITED, registered in England and Wales, registration number: 11086078, 5 Princes Gate Court, London, SW7 2QJ, United Kingdom

British Library Cataloguing-in-Publication Data
A catalogue record for this book is available from the British Library

Additional hard and PDF copies can be obtained from orders@intechopen.com

Avantgarde Reliability Implications in Civil Engineering
Edited by Maguid Hassan
p. cm.
Print ISBN 978-1-80355-102-9
Online ISBN 978-1-80355-103-6
eBook (PDF) ISBN 978-1-80355-104-3

# Meet the editor

Prof. Maguid Hassan earned his Ph.D. in Structural Reliability from the University of Maryland, College Park, USA. He participated in the establishment of civil engineering programs at several private higher education establishments in Egypt. He was awarded a Fulbright teaching/research grant in 2004, during which he taught structural engineering courses at the University of Maryland, College Park, and conducted research. Prof. Hassan joined the British University in Egypt (BUE) in 2006 to establish the Civil Engineering Department. In September 2013, he was appointed the associate dean for research, postgraduate studies, and community services. In November 2014, he was appointed the dean of engineering. Prof. Hassan's research interests include smart structural systems, structural health monitoring, structural reliability, structural steel analysis and design, applications of fuzzy logic and neural networks, and smart structural control.

# Contents

# Preface

Reliability and reliability-based design have emerged over the past several decades as integral and essential concepts in structural design. Engineering design problems are known to deal with several uncertainties; from material properties to load conditions, values, and combinations. Thus, the concept of 100% safe designs does not exist in real life. Therefore, it is only realistic to express all designs in terms of the lowest acceptable probability of failure, which results in acceptable reliability for these designs. This is referred to as a reliability-based design.

Another essential concept that emerged because of the increasing demand for sustainable design and performance of all engineering systems is structural resilience. The ability of any system to recover after a major event and perform satisfactorily is another major concern. Due to climate change, the noticeable increase in the severity and frequency of environmental events affecting civil systems such as buildings, roads, highways, and infrastructure dictated structural resilience and the response to all types of disasters, natural and/or those caused by humans, as essential factors in engineering design.

Recent advances in relevant technologies have led to the emergence of innovative solutions for the design of reliable, resilient, and sustainable designs. This book introduces several innovative solutions for achieving such designs. It is organized into four sections. The first section consists of an introductory chapter that presents a range of innovative solutions that are bound to represent the avantgarde direction of travel when dealing with reliability-based design. The chapter presents the impact of innovative concepts, in general, on the resulting reliability of engineering systems. Such concepts include sustainability, smart, health monitoring, disaster response, and resilience. The second section presents a state-of-the-art approach for achieving a reliability-based design. The third section highlights several artificial intelligence applications and their impact on the reliability of integrated systems. Finally, the fourth section presents several geotechnical applications and their impact on the reliability of resulting systems.

**Maguid Hassan**
Dean of Engineering,
The British University in Egypt,
Cairo, Egypt

Section 1

# Introduction

Chapter 1

# Introductory Chapter: Innovative Manifestations of Reliability-Based Design

*Maguid Hassan*

## 1. Introduction

Reliability and reliability-based design have emerged, over the past several decades, as integral and essential concepts in structural design. Engineering design problems are known to deal with several uncertainties, from material properties to load conditions, values, and combinations [1–3]. Thus, the concept of 100% safe designs does not exist in real life. Therefore, it is only realistic to express all designs in terms of the lowest acceptable probability of failure which results in acceptable reliability for these designs. That is what we all agree to refer to as reliability-based design [1–3].

Another essential concept emerged because of the increasing demand for sustainable design and performance of all engineering systems, which is structural resilience [4]. The ability of any system to recover after a major event and perform satisfactorily is another major concern. Due to climate change, the noticeable increase in the severity and frequency of environmental events affecting civil systems; buildings, roads, highways, and infrastructure, dictated that structural resilience and the response to all types of disasters, natural and/or man-made, emerged as essential factors in engineering design [5].

The attainment of a reliable and resilient design gave way to the emergence of innovative ideas in dealing with uncertain and continuously evolving design criteria. The recent interest in ensuring sustainability and sustainable design and the commitment by the United Nations in issuing its 17 Sustainable Development Goals (SDGs) would further strengthen the need to develop innovative design approaches that would satisfy all relevant concerns and criteria.

It is further important to emphasize that any such innovative approaches need to be reliability based to suite the nature of the highly uncertain problem that defines engineering design. As an example, the emergence of smart structural systems and the concept of structural health monitoring are meant to ensure the sustainable and reliable design of engineering systems.

It is essential though to indicate what does a sustainable design entail. Sustainability is a holistic concept that could be translated into several design criteria, such as reliable, smart, optimum, and recyclable [6]. In other words, any design approach that would result in less material use, optimum material use, recycling of material waste, reliable, and resilient system could be considered a sustainable design.

In this book, a range of innovative reliability-based solutions are presented to address the ever-changing engineering design problem. Such solutions range from


IntechOpen

considering stability issues of rock formations and thus their effect on the reliability of any proposed stabilizing system, to the issue of disaster response and how is that related to the resilience of engineering systems.

This introductory chapter will also attempt to present a range of innovative solutions that are bound to represent the Avantgarde direction of travel when dealing with reliability-based design. In the following discussion, the impact of the innovative concepts, in general, on the resulting reliability of engineering systems is discussed and portrayed. Such concepts include sustainability, smart, health monitoring, disaster response, and resilience.

## 2. Sustainability and reliability

Recent design criteria should not neglect an increasingly important factor that now, more than ever, is emerging as an integral design criterion. Sustainability and sustainable design are now considered to be one of the most important design criteria that needs to be integrated within all design activities [6]. Due to the climate change effects and the recent interest by the United Nations in devising the 17 Sustainable Development Goals, all designed systems shall ensure their sustainable performance throughout their expected lifetime [5]. Several sustainable goals are targeted when designing any of such systems, namely, goal 9 which relates to resilient infrastructure and goal 11 which relates to resilient and sustainable cities.

The main concept of sustainable design is one that should result in an environmentally friendly system. The concept of sustainable design, when considered, has always referred to the employment of recyclable material. In spite of using recyclable material being an important sustainability target, in its own right, yet there are other factors that would result in a sustainable design such as one that is optimum, smart, reliable, resilient, and green.

The most important design objectives are recyclable, optimum, and reliable. Thus, the smart characteristic of today's systems emerges as an important and critical attribute. In other words, the design of a system that is capable to adjust its physical properties to respond favorably to an unexpected event, natural and/or man-made, is much better, sustainable, and reliable than one that is designed to resist a load condition that it might not encounter during its lifetime. Therefore, smart systems are now considered a reliable sustainable option as compared to conventional systems.

It is important to emphasize that reliability emerges as a key factor in the design process; thus, the need for reliability-based design, yet such designs must be achieving advanced and continuously changing design criteria, such as sustainability. Another factor that has recently been identified as an important design objective, which is resilience, is due to the new approach toward performance-based design. Naturally, when considering a smart system that embodies several smart features that allow self-learning and continuous adaptation to uncertain events, this would result in a resilient system.

## 3. Smart systems and reliability

Smart structural systems are defined as ones that demonstrate the ability to modify their characteristics and/or properties to respond favorably to unexpected severe loading conditions. Conventional structural systems are usually designed to resist predefined loading conditions. However, due to the uncertain nature of engineering systems, and the lack of complete and accurate information about some types of highly uncertain loads, such as earthquakes, smart structural systems have emerged as a potential solution for such problems. Instead of designing systems to withstand a single extreme earthquake event that may or may not occur in its lifetime, new designs of smart systems could emerge where the system can respond favorably, in a smart manner, to any type of loading that was not introduced at the design stage [7]. The significance of such systems is even enhanced when modeled systems are unconventional such as historic buildings and/or structures.

The design of such systems requires the integration of sets of additional components within regular systems. These components belong to three major categories, sensors, processors, and actuators. The following section presents a summary about the nature of such components.

### 3.1 Basic components

For smart systems to behave as outlined above, they need to extend their capabilities through the employment of three basic component categories, i.e. sensors, processors, and actuators, [8, 9]. First, sensors are required to collect real-time data regarding the performance of the system in question. Acceleration and/or displacement transducers are considered as suitable devices for sensing any instantaneous state changes. Second, gathered real-time data are communicated to either a central processing unit or a set of decentralized processing units. The processing unit should be capable of identifying the current state of the system and, accordingly, suggesting corrective action if the response is beyond predefined and acceptable limits. Smart structural systems are designed to mimic the mechanical behavior of a human body. Therefore, any potential processor needs to possess cognitive features that are like those of the human brain. Such features include an auto-adaptive nature, parallel processing capability, and pattern recognition skills. Neural networks and/or fuzzy logic can simulate such cognitive features. Neural networks possess an adaptive self-learning capability, while fuzzy logic is capable of modeling complex systems that incorporate the qualitative nature of the human brain [10–12].

Finally, the processing unit sends a proposed corrective action to a set of actuators that follow one of two potential approaches. The first is an external force application to balance the system. The second is through the adjustment of the actuator's structural characteristics, thus, resulting in a more favorable response. External force application was one of the early forms of smart corrective actions [9, 13]. However, recent applications have shifted toward auto-adaptive actuators, which follow the second approach. Magneto-Rheological (MR) dampers are a good example of such actuators. MR dampers are proving to be very promising in civil and/or structural applications [14–16]. MR dampers respond to a magnetic field with a change in viscosity, thus, changing their response characteristics.

### 3.2 Reliability assessment framework

The main objective of this task is to outline a generic reliability assessment framework for evaluating the reliability of different types of components and ultimately the overall system reliability [17–21]. Being an integrated system, comprising a set of basic components, the individual component reliability is of major importance to the evaluation of the overall reliability of the system. As outlined above, smart structural systems require the integration of a set of sensors, at least a central processor unit and a set of actuators, in addition to several structural elements employed to build the system. By expressing such a system in a limit state format, one could easily employ standard algorithms in evaluating the reliability of each individual component and eventually the overall reliability of the system. The reliability assessment framework, for each type of integral component, necessary for the design of a smart structural system has been developed together with a sample limit state formulation that guide the specific development of individual component reliability schema. **Figures 1–4** show the developed reliability assessment frameworks for sensors, actuators, processors, and the overall system.

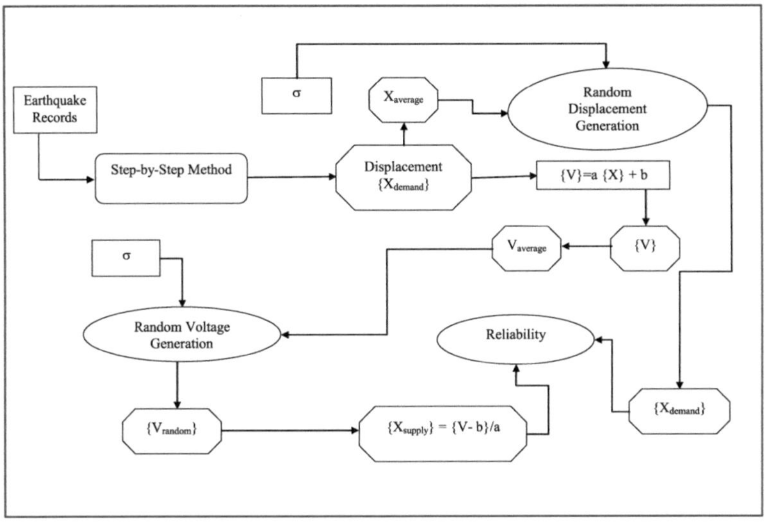


**Figure 1.**
*Sensor reliability framework.*

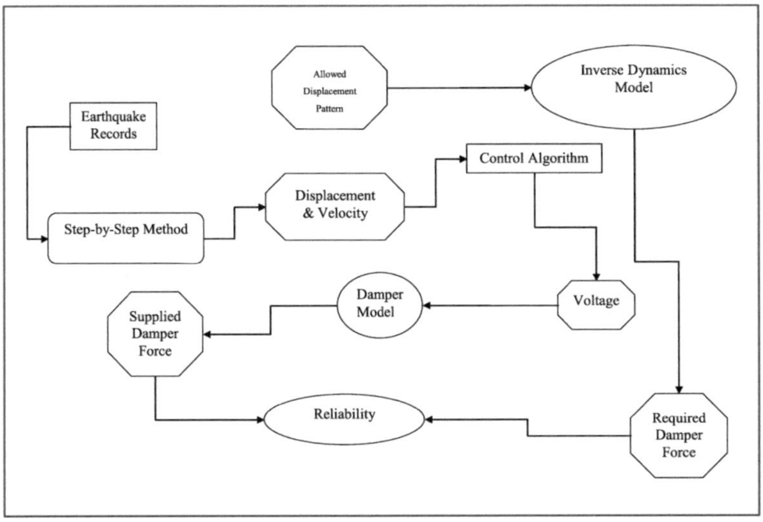


**Figure 2.**
*Actuator reliability framework.*

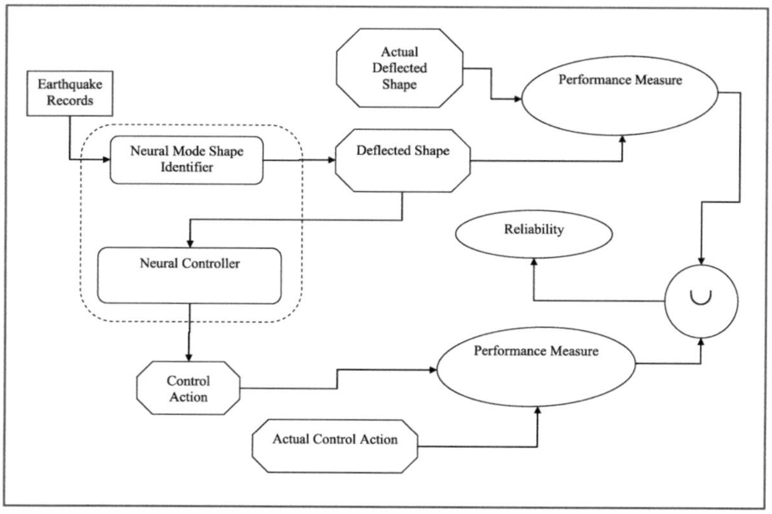


**Figure 3.**
*Processor reliability framework.*

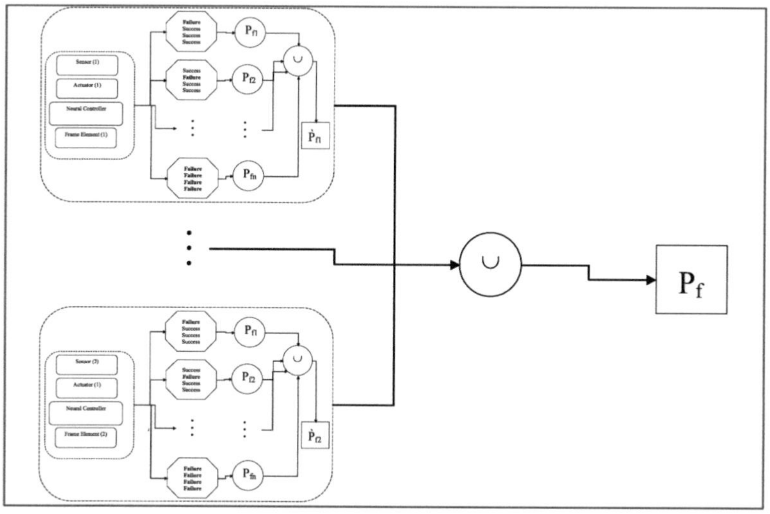


**Figure 4.**
*System reliability framework.*

## 4. Resilience and reliability

Disaster response is another major concern in the design of engineering systems. The ability of a system to recover after being exposed to a major event, whether that is a natural hazard and/or a man-made event, is one of the major concerns that engineering systems are now required to address [4, 5]. Resilience is known to reflect such ability, while reliability is meant to reflect the ability of such system to survive such events. Recent performance-based design approach led the way to further interest in structural resilience, which is concerned with such ability, at the component and system levels.

Resilience could be considered at three main levels, namely, structure and/or system level, infrastructure network level, and finally the community/urban level [5].

It is important to ensure the ability of systems, at all three levels, to recover and function satisfactorily after a major event.

At the structure and/or system level, this relates to structural systems and their redundancy which allow them to operate and function safely and reliably, even after such an event and until regain of full capacity is achieved. This is achieved through an acceptable level of redundancy that allows continuing function even after a local damage and/or failure occurs to one or more underlying components.

At the infrastructure level, this relates to the ability of highways, power grids, and all other utility networks to function and their inherent redundancy to ensure sufficient operation of such networks, if some damage is incurred and to ensure that such local damage does not propagate through the whole network and cause full disruption of services.

At the community and/or urban level, this relates to the socioeconomic performance of communities after a major disruptive event and the ability to recover and function normally.

The performance at these three levels is expected to be evaluated and expressed in terms of reliability measures since all such events are uncertain in nature. There are several attempts to evaluate and assess resilience measures that are available in the literature [4, 5, 22].

## 5. Health monitoring and reliability

Health monitoring relates to the continuous supervising of the performance of civil structures and systems throughout their lifetime. The advances in electronic devices, computer systems, and sensor technology paved the way for the development of integrated applications that can monitor the performance of civil systems, real time [23], and send alerts and alarming messages to the relevant authorities about potential damage and or recommended maintenance procedures.

Such a mechanism would enhance the reliability of targeted systems and ensures their continuous uninterrupted operation throughout their lifetime. However, the reliability of such systems is of major concern, since they comprise several components that are expected to operate integrally within a preset framework and work seemingly within a certain system [23].

The integration of structural health monitoring, within an engineering system, is expected to enhance its reliability. It is important to consider the reliability, of such system, in two main levels, the first is the reliability of the structural health monitoring system itself and the second is the reliability of the monitored engineering system, including the health monitoring enhancement. There is not enough research activity in reliability of structural health monitoring systems and their impact on the reliability of monitored systems. This requires a targeted research effort in this area which would support the practical implementation of such systems, especially when dealing with the conservation of historic cultural heritage.

## Author details

Maguid Hassan
The British University in Egypt, Cairo, Egypt

*Address all correspondence to: mhassan@bue.edu.eg

## References

[1] Haldar A, Mahadevan S. Probability, Reliability and Statistical Methods in Engineering Design. New York: John Wiley & Sons, Inc; 2000

[2] Nowak AS, Collins KR. Reliability of Structures. Boston: McGraw-Hill Inc; 2000

[3] Ang AH-S, Tang WH. Probability Concepts in Engineering Emphasis on Applications to Civil and Environmental Engineering. New York: John Wiley & Sons, Inc; 2007

[4] Lim S, Kim T, Song J. System-reliability-based disaster resilience analysis: Framework and applications to structural systems. Structural Safety. 2022;**96**:102202

[5] Corotis RB. Planning for community resilience under climate uncertainty. In: Murphy C, Gardoni P, McKim R, editors. Climate Change and Its Impacts: Climate Change Management. Cham: Springer; 2018. DOI: 10.1007/978-3-319-77544-9_9

[6] Hassan MHM. Fuzzy controllers: A reliable component of smart sustainable structural systems. In: Iqbal S, editor. Fuzzy Controllers-Recent Advances in Theory and Applications. 2012. ISBN: 978-953-51-0759-0

[7] Hassan MHM. A system model for reliability assessment of smart structural systems. Structural Engineering & Mechanics, An International Journal. 2006;**23**(5):455-468

[8] Spencer BF Jr, Sain MK. Controlling buildings: A new frontier in feedback. Control Systems Magazine, IEEE. 1998;**6**:19-35

[9] Soong TT. Active Structural Control, Theory & Practice. New York: John Wiley & Sons Inc; 1990

[10] Casciati F, Faravelli L, Yao T. Application of fuzzy logic to active structural control. In: McDonach A, Gardiner PT, McEwen RS, Culshaw B, editors. Proceedings of the Second European Conference on Smart Structures and Materials. Glasgow, Scotland. 1994. pp. 206-209

[11] Faravelli L, Venini P. Active structural control by neural networks. Journal of Structural Control. 1994;**1**(1-2):79-101

[12] Faravelli L, Yao T. Use of adaptive networks in fuzzy control of civil structures. Microcomputers in Civil Engineering. 1996;**11**:67-76

[13] Connor JJ. Introduction to Structural Motion Control. New Jersey: Prentice Hall, Pearson Education; 2003

[14] Spencer BF Jr, Dyke SJ, Sain MK, Carlson JD. Phenomenological model for magneto-rheological dampers. Journal of Engineering and Mechanical. 1997;**123**(3):230-238

[15] Carlson JD, Spencer BF Jr. Magneto-rheological fluid dampers for semi-active seismic control. In: Proceedings of the 3rd International Conference on Motion and Vibration Control. Chiba, Japan; 1996. pp. 35-40

[16] Dyke SJ, Spencer BF Jr, Sain MK, Carlson JD. Modeling and control of magneto-rheological dampers for seismic response reduction. Smart Material Structure. 1996;**5**:565-575

[17] Hassan MHM. Reliability evaluation of smart structural systems. In: IMECE 2005, ASME International Mechanical Engineering Congress & Exposition. Orlando, Florida, USA; 2005

[18] Hassan MHM. Reliable Smart Structural Control. 5th European Conference on Structural Control, EACS 2012. Genoa, Italy; 2012

[19] Hassan MHM. Real-time smart shape identifiers. In: 4th International Conference on Smart Materials, Strutcures & Systems, CIMTEC 2012. Montecantini Terme, Italy; 2012

[20] Hassan MHM. Toward reliability-based design of smart pattern identifiers for semi active control applications. In: Fifth World Conference on Structural Control and Monitoring, 5WCSCM 2010. Tokyo, Japan; 2010

[21] Hassan MHM. A reliability assessment model for MR damper components within a structural control scheme. Advances in Science & Technology. 2008;**56**:218-224

[22] Henry D, Ramirez-Marquez JE. Generic metrics and quantitative approaches for systems resilience as a function of time. Reliability Engineering and System Safety. 2012;**99**:114-122

[23] Hassan MHM, editor. Advances in Structural Health Monitoring. Open Access Peer-Reviewed Volume. IntechOpen; May 2019

Section 2

# Reliability-Based Design

Chapter 2

# An Efficient Quantile-Based Adaptive Sampling RBDO with Shifting Constraint Strategy

*Shima Rahmani, Elyas Fadakar and Masoud Ebrahimi*


## Abstract

There is an increasing demand for the performance optimization under the reliability constraints in various engineering problems. These problems are commonly known as reliability-based design optimization (RBDO) problems. Among different RBDO frameworks, the decoupled methods are widely accepted for their high efficiency and stability. However, when facing problems with high nonlinearity and nonnormally distributed random variables, they lose their computational performance. In this study, a new efficient decoupled method with two level quantile-based sampling strategy is presented. The strategies introduced for two level sampling followed by information reuse of nearby designs are intended to enhance the sampling from failure region, thus reducing the number of samples to improve the efficiency of sampling-based methods. Compared with the existing methods which decouples RBDO in the design space and thus need to struggle with searching for most probable point (MPP), the proposed method decouples RBDO in the probability space to further make beneficial use of an efficient optimal shifting value search strategy to reach an optimal design in less iterations. By comparing the proposed method with crude MCS and other sampling-based methods through benchmark examples, our proposed method proved to be competitive in dramatically saving the computational cost.


**Keywords:** reliability-based design optimization (RBDO), decoupled method, Stochastic optimization, Monte Carlo simulation, data-driven modeling, information reuse

## 1. Introduction

It is a high priority for the engineers to reach an optimal design scheme for the entire system. However, optimality alone cannot guarantee the reliability of the design since theoretically, the optimal design solutions are very close to the boundaries of the reliability constraints. So, the optimal solution in practice can be very sensitive to the design uncertainties and may easily crosses the reliability constraint limits. To avoid such a situation, the optimal design should either be represented with a confidence range (margin) or an optimization process with the objective of maintaining the desired level of reliability [1]. Reliability-based design optimization

(RBDO) methodologies have been introduced to incorporate uncertainties in the design components and to provide a scheme seeking an optimal solution with the desired level of reliability.

To better understand the mechanism of these methods, researchers divided them into three major categories [2]:

1. ***Double-loop strategies:*** This model is not very efficient as it accommodates nested optimization loops where the inner loop is the reliability analysis and the outer loop runs the main deterministic design optimization (DDO) problem subject to the reliability requirements [3]. The computational inefficiency lies in the fact that each constraint is under the reliability analysis even if it is very far from the deterministic optimal point and thus has a confidence level (probability of success) very close to 1 [4].

2. ***Decoupled strategies:*** If the reliability analysis is not called at every cycle of the main optimization, it is classified as a decoupled strategy [5].

    - ***Unilevel methods:*** It replaces the inner reliability analysis by Karush–Kuhn–Tucker (KKT) optimality conditions on the optimum point and then enforces it as a constraint itself in the main design optimization loop [2, 5–7].

    - ***Single loop methods:*** This approach arranges the reliability analysis and DDO in a sequential manner so that in each cycle, the probabilistic constraints are transformed into equivalent deterministic constraints [8–13].

The failure probability estimation is both the key component and challenge of an RBDO problem.

Three types of methods have been identified to calculate the failure probability of constraints [14]:

1. ***Analytical methods:*** The concept is that if the most probable point (MPP) after being found, satisfies the probabilistic constraint, then the design can be considered reliable. Methods for reliability assessment involve reliability index analysis (RIA) and performance measure analysis (PMA) in the frame of first-order reliability method (FORM) and second-order reliability method (SORM) [15, 16]

2. ***Numerical integration***: In this method, a distribution function generates an approximate density function of the system response based on statistical moment estimates which allows for the estimation of failure level (e.g., dimension reduction methods, eigenvector dimension reduction methods and multi-dimensional integration form [17–19]).

3. ***Simulation-based methods***: This method with its primary form "crude MCS" could result in high accuracy of estimation if a sufficiently large sample size is used. However, this method can often become computationally prohibitive, and hence, recent sampling methods have been proposed to overcome this obstacle. To cite a few, importance sampling (IS), subset simulation (SS) and line sampling are the most frequent approaches in this regard [20–30]. To help with the heavy computational load of simulation-based methods especially for small failure

probabilities, an inexpensive surrogate model is employed to approximate the original high-fidelity model. These methods are classified into three main groups:

- ***Data-driven methods*** (i.e., artificial neural network [31], kriging [32–34], polynomial chaos expansion (PCE) [35], response surface method (RSM) [36, 37], support vector machine [38])

- ***Hierarchical-based methods*** [39]

- ***Projection methods*** [40]

If the constraints are mildly nonlinear, analytical methods (approximate function methods like FORM and SORM) as well as surrogate methods (e.g., RSM, Kriging) will provide rather a good estimation of the reliability. In contrast, simulation-based approaches have the advantage of not being sensitive to the complexity of the limit state functions (LSF) and avoiding LSF approximation errors [41]. The high expenses of MCS in the presence of expensive LSF, however, make their combined usage inefficient. Even the advent of surrogate models for coping with this issue have proven to cause a bias in failure probability estimation. They also do not reuse data from the previous design iterations which can save a great amount of computational budget [30]. Other approaches for decreasing the computational efforts of simulation-based methods, especially when encountering smaller failure probabilities, are IS and subset sampling. One of the major disadvantages of these sampling techniques is their need to design an initial proposed distribution. If the biasing density is not built appropriately, it can be problematic for high-dimensional and highly-nonlinear LSF since they still require a huge sampling for the estimations.

In RBDO, both the reliability assessment and its integration with the optimization process to find not only an optimal design but also a reliable one are vital. To make a balance between computational efficiency and accuracy of reliability estimate, decoupled methods proved to be good choices. They break down the nested loop structure of DDO and the reliability analysis and perform them sequentially. Sequential optimization and reliability analysis (SORA) [4] is the most popular method in literature with a desirable stability and efficiency [42] in which the feasibility of the current optimal point for probabilistic constraints is assessed by the MPP [43]. However, it has the shortcoming of MPP-based methods discussed earlier. In recent years, most of the studies in SORA have concentrated on addressing how to efficiently search the MPP in a decoupling procedure [42].

The main contribution of this work is twofold:

- integration of the concepts of SORA's shifting constraint strategy with the calculation of quantile-based probability of failure,

- using a combination of recycled samples and newly generated samples in each iteration to make use of computational wastes.

The first idea aims at eliminating the need to search for MPP and thus is able to overcome the inherent limits of the MPP-based methods. To make this adjustment, unlike SORA which decouples RBDO structure in the design variable space, the proposed method decouples RBDO in the probability space. By calculating the required distance to reach a more reliable design in the probability space, the reliability constraint is converted into a deterministic constraint. Then, the RBDO problem is defined as a sequence of sub-deterministic optimization problems.

The second proposition aims to improve the computational efficiency. For the failure probability calculation, the quantile is computed by the use of sampling methods. In order to reduce the computational effort for sampling, two strategies are proposed in each iteration; firstly, adaptive constraint shifting based on the required level of reliability in each iteration is suggested, leading to a significant reduction in iterations. Secondly, once the initial sampling for forming the constraint distribution around the current optimal point is performed, new samples are generated. To generate the samples, statistical properties of infeasible region of constraint distribution are used. In the meanwhile, the samples generated from previous designs are recycled to avail the current iteration. A mixture of these recycled samples with newly generated samples will be further used in the reliability analysis to provide an accurate estimate of quantile. By doing so, the convergence to the final reliable design point is gained.

To the best of our knowledge, such an idea has not been implemented under quantile-based RBDO concepts. Reusing existing information in the context of RBDO has been identified in [30, 44]; however, they used them in a double-loop RBDO framework which suffers from heavy computational burden discussed earlier. The idea of recycling samples comes from the assumption that the nearby designs are likely to have similar failure regions. Hence, reusing them can result in computational saving that is required for a better description of the failure region.

The remainder of the paper is organized as follows:

The proposed framework with its innovative adjustments is described in detail under Section 2. Section 3 studies the performance of the proposed method by applying it on a number of highly cited examples in literature that is followed by a discussion on its advantages. At the end, the key characteristics of the proposed method as well as the conclusion of the achieved results are summarized.

## 2. Proposed sampling-based RBDO framework

SORA is dependent on the location of MPP to find the shifting vector. The strategies of searching MPP are always based on the gradient methods which may drop into the local optimum or oscillate between multiple MPPs. Another error arises when random variables are not normal since they have to be mapped into equivalent standard normal space [42]. In this study, the sensitivity of the failure probability to solve the RBDO problem is based on the adaptive sampling without any need to map the random variables to the standard normal space during the MPP-search procedure. In this framework, constraint shifting is carried out in the probability space.

### 2.1 RBDO problem formulation

A generic formulation of the RBDO problem is expressed as

$$\begin{aligned}
&\mathit{Minimize}: \quad f\left(\boldsymbol{d}, \mu_X, \mu_p\right) \\
&w.r.t.: \quad Prob\left[g_i(\boldsymbol{d}, X, P) \geq 0\right] < P_f^t, \qquad i = 1, 2, \ldots, n \\
&\boldsymbol{d}^L \leq \boldsymbol{d} \leq \boldsymbol{d}^U, \quad \mu_X^L \leq \mu_X \leq \mu_X^U, \quad \mu_P^L \leq \mu_P \leq \mu_P^U
\end{aligned} \tag{1}$$

*DOI: http://dx.doi.org/10.5772/intechopen.110442*

Where **d** is a deterministic design variable, $X$ is a stochastic design variable vector, and $P$ is an uncertain design parameter vector. The inequality function, $g_i(\boldsymbol{d}, X, P) \geq 0$, is generally the safety requirement, where $g_i > 0$ indicates the safe region, $g_i < 0$ indicates the failure (infeasible) region, and $g_i = 0$ defines the limit state surface. The value $P_f^t$ is the target failure probability, so for the probabilistic constraint to guarantee the design's reliability, this value should be reached.

An equivalent model to Eq. (1) aligned with quantile-based probability estimation is given by

$$\begin{aligned} &Minimize: \quad f\left(\boldsymbol{d}, \mu_X, \mu_p\right) \\ &w.r.t.: \quad g_i^R(\boldsymbol{d}, X, P) \leq 0, \quad i = 1, 2, \ldots, n \\ &Prob\left[g_i(\boldsymbol{d}, X, P) \geq g^R\right] = P_f^t, \end{aligned} \tag{2}$$

Which indicates that the probability of $g_i(\boldsymbol{d}, X, P)$. is greater than or equal to R-percentile $g^R$ which is exactly equal to the target failure probability $P_f^t$.

To better understand the shifting of unreliable constraints toward safe region in a probability space (rather than random variable space), the RBDO problem is formulated as

$$\begin{aligned} &Minimize: \quad \mathbb{E}[f(\boldsymbol{d}, X)] \\ &w.r.t.\ Prob\left(g_i(\boldsymbol{d}, X) < 0\right) \leq P_f^t \end{aligned} \tag{3}$$

For simplicity in notations, $X$ denotes all random samples (whether design variables or design parameters).

In each iteration, once the DDO is solved, the reliability of constraints will be evaluated on the current optimal design point. A shifting vector is required to be searched in the probability space which implies the required step size for moving the boundaries of constraints toward the feasible region for the next iteration. This will ensure that the optimal point gained from the deterministic optimization will be located on the deterministic boundary of constraints with improved reliability level. In this framework, efforts are dedicated to seek for the best shifting value for each active constraint in order to efficiently shift the probabilistic constraint toward the feasible region. Moreover, it should be emphasized that the distance between the quantiles of $P_f^t$ and the optimal point of current design is measured in the probability space contrary to the SORA that measures the distance between MPP and the optimal point in a standardized design space. Doing so increases the stability and accuracy of the method when solving the RBDO problem. This will be illustrated through the examples section.

As in **Figure 1**, if the value of $P_f^t$ is predefined, by shifting the limit state $(g = 0)$ from its original place to the location of line 2, the target failure probability will be obtained. This shift for limit state line can be done by means of an appropriate auxiliary shifting variable "$y$".

The minimum value of $y$. is zero and begins from where the constraint $g$ is zero $(g = 0)$. On the other hand, $y$ can take a maximum value equal to the maximum value of $g$ or the endpoint of the tail of PDF termed $UB_g$.

To calculate the required shifting for constraints, a very simple line search problem should be defined as

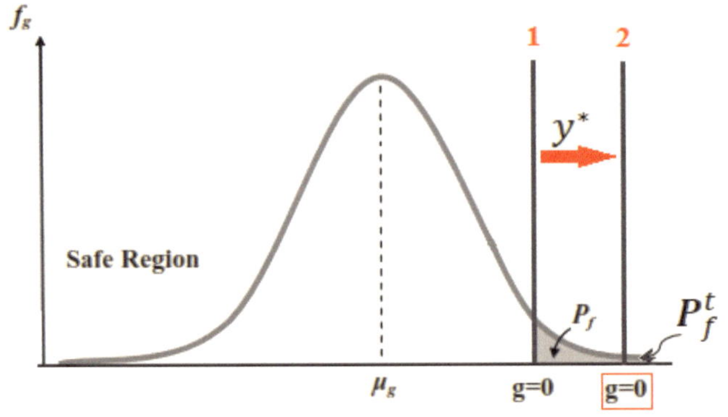


**Figure 1.**
*Shifting the limit state line as much as an optimal value $y^*$ to satisfy the desired reliability ($P_f^t$) for constraint g in probability space.*

$$\begin{aligned} &Minimize \quad -y \\ &s.t. \ P_f - P_f^t \le 0 \\ &0 < y < UB_g \end{aligned} \tag{4}$$

Where the shifting variable "$y$" can only take values between 0 and upper bound of constraint distribution "$g$". The amount of "$y$" should be maximized while the probability of failure is less than target failure probability. It states that seeking for a "$y$" is aimed so that the percentage of samples (of $g$'s distribution) that violates the value of "$y$" becomes equal to the target failure probability.

As it is stated in [4], as the design moves toward 100% reliability, the computational effort for searching the MPP in a standard normal space becomes greater. However, in the proposed approach, part of this issue is solved in the reliability assessment level by retrieving the useful historical data from nearby designs, and part of it can be addressed by relying on an adaptive "$y$" search strategy.

## 2.2 Adaptive shifting value search

If the value of $y$ is chosen to be a very small amount, then the constraint will be shifted at a very low pace. Subsequently, the rate of convergence will decrease, and hence, after a great number of iterations the probability requirements would be met. On the other hand, if $y$ takes a large value, it is very likely to have a premature design with higher reliability than is desired, meaning that an overdesign has occurred. To tackle this problem, an adaptive step size is introduced so as to make a tradeoff between the amount of constraint shifting and the level of required $P_f$ . The essence is to move the design solution as quickly as possible to its optimum in order to eliminate the need for locating MPP that is adjusted to simulation-based approaches. The tradeoff problem is formulated as below:

$$Cost\ function = y_{ratio} + \Delta P_f \tag{5}$$

$$y_{ratio} = \frac{y}{UB_g} \tag{6}$$

$$\Delta P_f = \left( \frac{1}{N} \sum_{i=1}^{N} \mathbb{I}_{inf}(d, x_i) \right) > y \tag{7}$$

Where the indicator function $\mathbb{I}_{inf}$ is defined as

$$\mathbb{I}_{inf}(d, X) = \begin{cases} 1, & x \in inf\ , \\ 0, & \text{else.} \end{cases}$$

And $x_i, i = 1, \dots, N$ are the $N$ samples from probability distribution of constraint $g$. The first term in Eq. (5) denotes the normalized shifting value, and the second term denotes the error in failure probability.

The above problem can be stated in the form of a penalty function:

$$Cost\ function = y_{ratio} + r\left[\sum (\max\{0, v\})^2\right] \tag{8}$$

and

$$v = \left[ \left( \frac{1}{N} \sum_{i=1}^{N} \mathbb{I}_{inf}(d, x_i) \right) > y \right] - P_f^t \tag{9}$$

Where multiplier "$r$" determines the severity of the punishment and must be specified a priori depending on the type of the problem (e.g., small failure problems).

The main advantage of using this penalty function is that in the initial iterations (less reliable designs), the shifting toward safe region would be quicker by putting more weight on "$y$", whereas by approaching to the last iterations where the infeasible region becomes narrower, the second term of Eq. (5) outweighs the first term. Consequently, it does not allow for taking big step sizes and exceeding the $P_f^t$ at later stages especially for rare events.

It should be noted that in Eq. (8), "$r$" plays the role of a controller to control the effect of the error term which should be changed adaptively. For instance, it can be stipulated that when the difference between $P_f$ and $P_f^t$ is less than an adequately small amount (e.g., 0.1), then an increase in the value of "$r$" will be applied, that is, magnifying the error in the failure probability. This is especially beneficial for the problems with stricter reliability requirements.

We use the results of the example 3.1 (section3) in advance for intuition.

The value of "$r$" in the following problem is set to 1e+3 since the $P_f^t$ is 1e−3. Accordingly, if the $P_f^t$ was 1e-4, "$r$" should be set to 1e+4. To enlighten how the formula works, the tradeoff curve between the amount of shifting and the probability failure error is provided in **Figure 2**.

As is evident in **Figure 2**, by employing such optimization formulation, a tradeoff between two terms of the objective function is made; first, greater values of "normalized $y$" is in favor of minimizing the penalty function, because the great amount of failure probability has more impact on the penalty function and tries to make it maximum as fast as possible. Then, the trend will be reversed, meaning that the amount of shift becomes more important. Narrower failure region in the next iterations does not cause the effect of "$y$" to fade since "$y$" is normalized. The second term is squared so as not to allow any violation from $P_f^t$ (occurrence of overdesign).

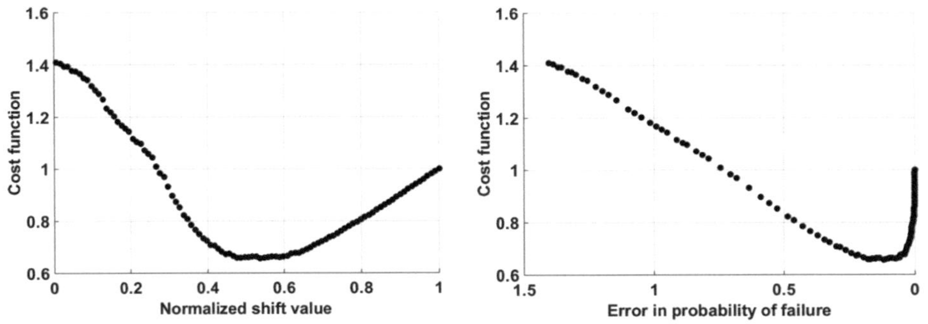


**Figure 2.**
*Results of the adaptive optimal shift value search for example 1. The left panel indicates the effect of shift value step size on the cost function, and right panel indicates a sharp impact of error in the probability of failure on the cost function.*

This shifting value search algorithm is robust because it is suitable for both continuous and discrete constraint functions.

Defining $y$ and solving its optimization problem will not impose a significant computational load, because in fact, a comparison between a predefined value ($P_f^t$) and available samples is performed. As such, in each iteration a value for $y$ is suggested which will be added to the deterministic constraint of the next iteration. Then, the DDO problem aiming at finding new optimal solution will be done considering the shifted constraint in a decoupled DDO-reliability assessment approach.

The algorithm will be in progress until a stopping criterion is satisfied. A proper stopping criterion for this framework could be a maximum shifting value or shifting vector (in case of more than one constraint). This criterion implies that constraints are all reliably satisfied. Therefore, by this search strategy adapted to simulation-based RBDO, the reliability of constraints improves progressively, while the need for searching MPPs is eliminated.

### 2.3 Sampling strategy for reliability analysis

To get an insight into the constraint's behavior near the failure region, the first level of sampling (L1) is required. This will happen around the optimal point identified in the DDO of current iteration with the distribution properties defined for the random variables. Contrary to the existing sampling-based methods [30, 44], in the proposed method, for a more efficient sampling with least wastes, central limit theory is used. According to this theory, "the mean of a batch of samples will be closer to the mean of the overall population as the sample size increases" [45]. The overall population here means the huge MCS samples required to describe the probability distribution of the active constraints. Therefore, an adaptive sampling can be employed to distribute the initial samples around the optimal point of the current iteration regarding the difference between mean of the batch of samples and the mean of the whole population. By the aid of such adaptive sampling in each iteration, a more appropriate generation of sample size with less waste can be obtained. Moreover, this sampling will provide a distribution function for the constraint that has a better description of the failure region and hence provides a proper basis for the second-level sampling. The second-level sampling (L2) aims at distributing more samples with a focus on the region significant to the RBDO.

MCS has been run 500 times for example 1 (Section 3) to examine the process of convergence to the population mean (**Figure 3**). Then, it is repeated 100 times to

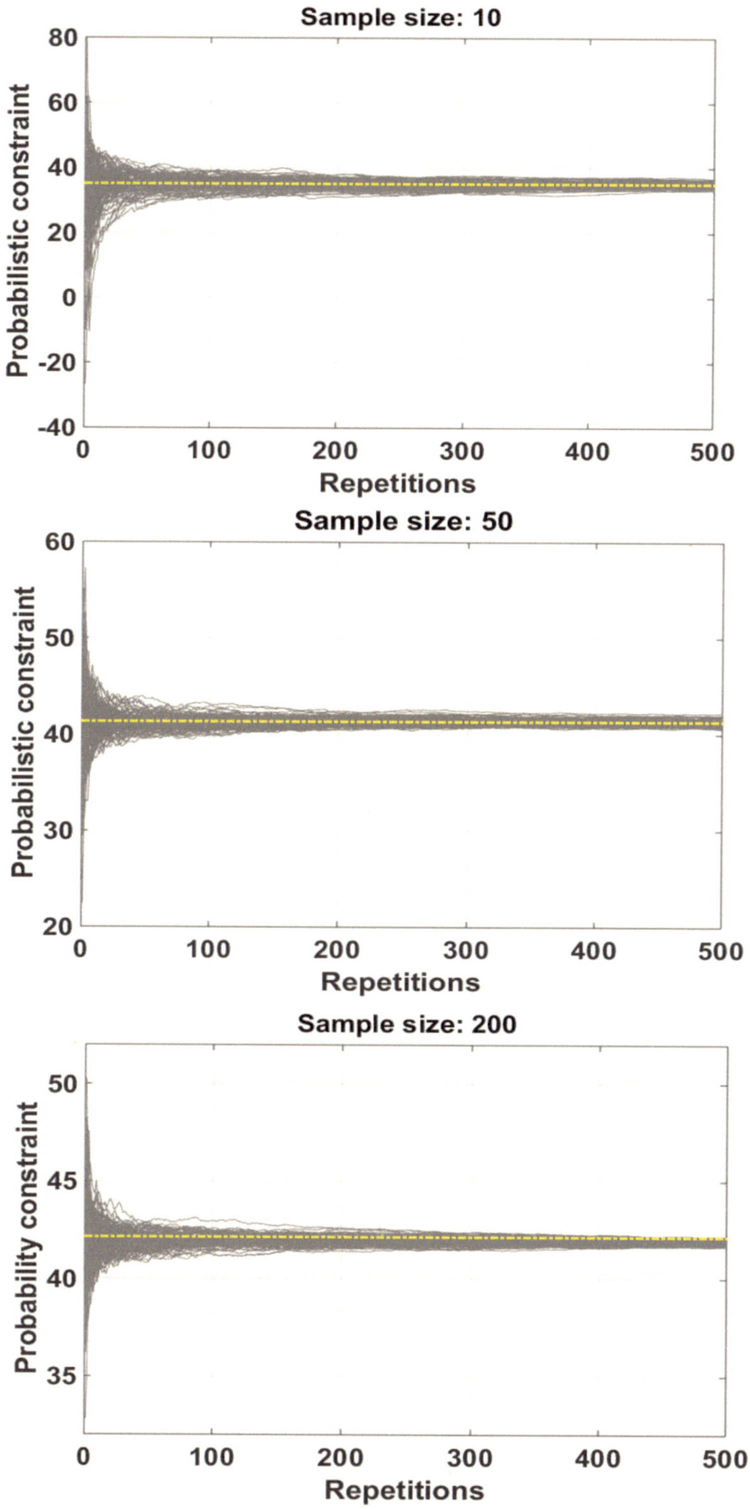


**Figure 3.**
*Variability for different sample sizes in 500 repetitions of MCS.*

exhibit the variability. In each run, a specified number of samples is generated, and then, these samples are aggregated with the previously produced samples. Then, the mean of the batches of samples is plotted in order to trace the value to which their means are approaching. Three different sample sizes are depicted to show the amount of variability for each one of them.

Across the 500 repetitions, there was noticeable variability at low sample sizes. As expected, this variability decreases with larger sample sizes. Thus, in each RBDO iteration, an average sample size can be considered. Then, as long as the coefficient of variation (CoV) is not reached its desirable value, the process of generating the batch of samples will continue. In example 1 (Section 3), the sample size has been set to 50. The total number of samples generated to obtain a CoV of 0.1 was 1000 for the first iteration. It is clear that as the iterations continue, due to narrower failure region, more data is required. This volume of samples will be then adaptively determined based on the preset CoV. The impact of using CoV as the stopping criteria rule depends on the properties of the probability space. This adaptive initial sampling also reduces the potential error of improper sample size selection when building biasing density functions in IS methods.

Now, the generated samples will be used both for estimating the failure probability and the second level of sampling. In order to more accurately identify the constraint's quantile of each cycle, two actions are taken; (1) Generating second level of sampling with its focus on the failure region, and (2) Recycling the historical samples (samples from previous iterations). For quantile-based failure probability, only the samples generated in the tail of the constraints are significant to an RBDO problem.

It goes without saying that initial sampling around the optimal point results in a number of samples dropped in the safe region and the rest dropped in the failure region. The latter are those carrying useful information for second level sampling.

It should be noted that once initial sampling is carried out, inactive constraints can be identified to avoid their contribution in second sampling. The strategy proposed by [46] can be employed to remove inactive constraints regardless of having knowledge about their boundaries.

**Figure 4** (top) shows the 100 random data generated from a normal distribution with mean of 0.87 and standard deviation of 0.3 in the L1. The location of $x_{mpp}$ is indicated as well. The panel has broken and classified the random samples to two parts: pre-$x_{mpp}$ and post-$x_{mpp}$ as are demonstrated in **Figure 4** (bottom). So, there are two subsets of the main population of L1-sampling. It should be noted that regardless of the distribution types of random variables (i.e., normal, gumbel, weibull, etc.), to produce new samples, it is possible to safely use normal distributions. This is allowed by the central limit theorem [45].

The standard deviation of the sampling distribution is smaller than the standard deviation of the population by a factor of $\sqrt{n}$ [45], and hence, with this distribution, most of the L2-samples are distributed around the infeasible region.

Sorting the generated samples of random variables based on the index $g = 0$ of constraint distribution gives the location of $x_{mpp}$ in the probability space.

Magnification of the infeasible region is subject to sufficient sampling from this region. Thus, new random samples will be generated with the mean and standard deviation of L1-samples dropped into the infeasible region (**Figure 4**, bottom right panel).

In failure probability estimation with IS, the smaller the variance of constructed biasing density, the more accurate the failure estimation. Chaudhuri et al. [30] have

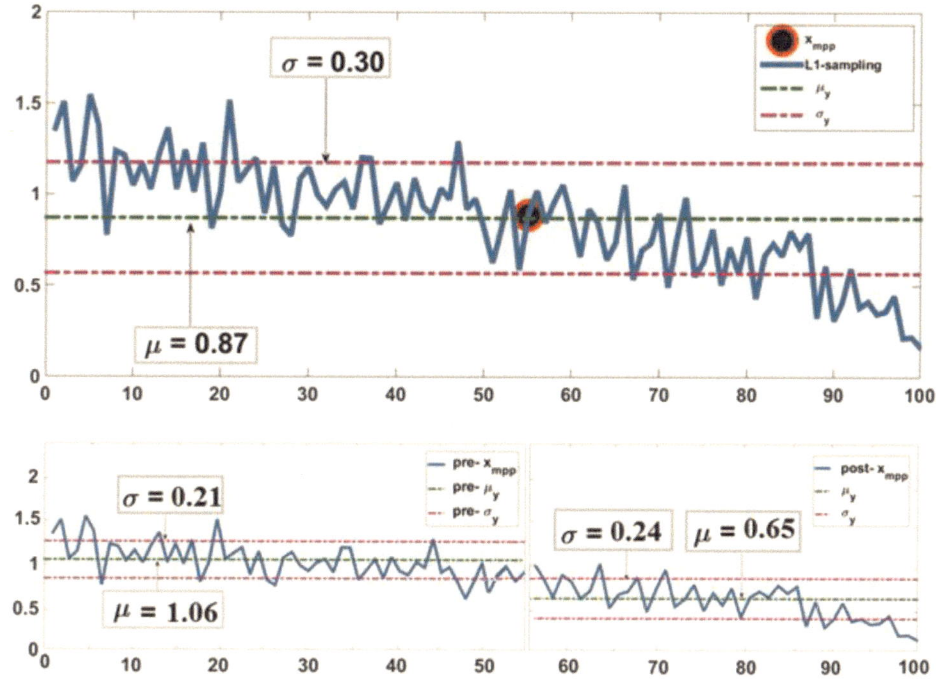


**Figure 4.**
*Top panel: Sample points generated for level one with a sample size of 100. Bottom panels: The same samples of upper panel but broken from the $x_{mpp}$ point into two separate panels each with its own mean and standard deviation.*

utilized the information reuse when building a posteriori biasing density to reach an optimal biasing density. Then, they sum the currently built biasing density with biasing densities of nearby designs to form the final density that will be used for failure probability estimation. In our proposed method, such strategy is employed but in a quantile-based frame to produce an accurate estimation of the failure probability. The newly generated samples along with the samples of the nearby design that have an overlap with the tail of current constraint's distribution will form the final distribution function. Our method is different from [30] in the way the failure probability is calculated. In quantile-based method, the final distribution of the constraints will be directly used in the stage of searching for an optimal shifting value that is intended to reduce the failure probability error. In the implementation of the information reuse, no evaluation of the expensive LSF is required.

The new distribution function will be as below

$$g_f = g + g_{inf} + \sum_{j=0}^{k-1} \acute{g}_j \tag{10}$$

Where $g$ is the initial distribution (from L-1 sampling), $g_{inf}$ is the distribution of L2-sampling, and $\acute{g}_j$ is the distribution built over all contributing samples from past RBDO iterations $j \in \{0, \dots, k-1\}$ that are stored in a database.

Similar to IS method where each sample has its own weight, in quantile-based method, each sample added to the $g$ distribution causes a change in the MCS estimator. That is because by adding any single sample, the denominator in Eq. (7) will increase [42]. Thus, using the mixture distribution in the process of optimal shifting value search plays an important role in the accuracy of estimations.

Considering the recycled samples, L1-sampling and L2-sampling, the problem can be reformulated as

$$\Delta P_f = \left( \frac{1}{n_{L1} + n_{L2} + m} \sum_{i=1}^{n_{L1}+n_{L2}+m} \mathbb{I}_{inf}(d, x_i) \right) > y \tag{11}$$

Where the indicator function $\mathbb{I}_{inf}$ is defined as:

$$\mathbb{I}_{inf}(d, X) = \begin{cases} 1, & x \in inf\,, \\ 0, & \text{else.} \end{cases}$$

Here $inf$ stands for the infeasible region of $g_f$, $n_{L1}, n_{L2}, m$ are the samples from $g$, $g_{inf}$ and $\acute{g}_j$, respectively.

**Figure 5** shows the flowchart of the proposed algorithm.

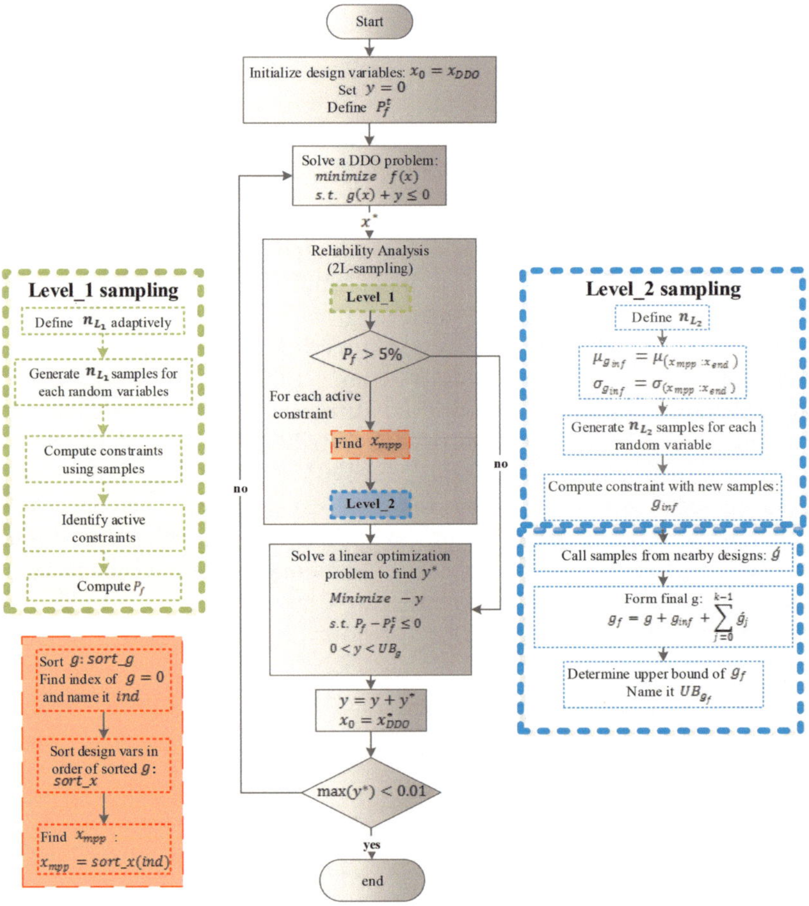


**Figure 5.**
*Flowchart of the proposed algorithm.*

In the next section, three well-known benchmark problems are employed to verify the efficiency and/or accuracy of the proposed method and to examine the whole proposed RBDO framework for quantile-based sampling methods.

## 3. Implementation and performance evaluation

In this section, the performance of the proposed framework in terms of applicability and computational efficiency will be investigated through three examples which are among widely cited benchmark problems. Then, the results will be compared with those previously presented for the sake of verification. To solve the deterministic optimization part, the sequential quadratic programming (SQP) through *fminbnd* function in MATLAB is used.

Since the computational cost of the deterministic optimization is negligible for sequential RBDO methods, the number of LSF evaluation (short form: NFE) is a fair measurement for efficiency.

The RBDO optimal solutions will be validated by using a crude MCS of ten million-size as a reference to satisfy the target failure probability.

### 3.1 First RBDO problem: buckling of a straight column

A structural engineering problem is considered to practice both the efficiency and accuracy of the method. A long straight column with a rectangular cross section, fixed at one end and free at the other is considered (**Figure 6**) [47]. A deterministic axial load F is applied at the free end.

The column design optimization is defined so that the total area of the cross section should be minimized while the exerted load should not exceed the critical buckling load. This is formulated as below:

$$\begin{aligned} &Minimize \;\; f(x) = bh \\ &s.t. \;\; P_f(x) \le P_f^t \\ &b - h \ge 0 \\ &g(x,p) = \frac{\pi^2 E b h^3}{12L^2} - F_a < 0 \end{aligned} \tag{12}$$

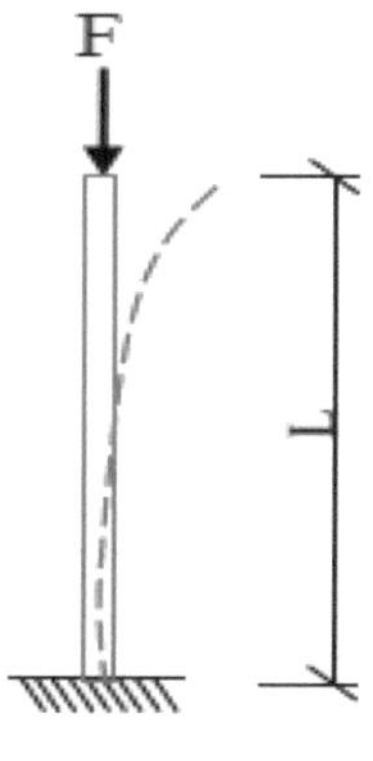


**Figure 6.**
*A straight column with an axial load applied on it.*

Where $b$,$h$ and $p$ denote the width, the height of the column and random parameters, respectively. $h, b$ are chosen to be the design variables.$E$, $L$ and $F$ stand for elasticity modulus, length of the column and axial load. Distribution information of $E$ and $L$ as random parameters and lower/upper bounds of $b$ and $h$ as deterministic variables are given in **Table 1**.

To examine the ability of the proposed algorithm to reach different target failure probabilities, three cases are considered (**Table 2**).

The results of **Table 2** indicate that our algorithm not only is capable to satisfy the desired level of reliability but also compared to the other decoupled sampling-based method, uses less function calls. The algorithm has presented an efficiency of around 9, 5 and 2 times better in the first, second and third cases, respectively, in comparison with the other sampling method.

The probability of failure for the probabilistic constraint of each case at the optimum is evaluated by crude MCS with a ten-million sample points so as to ensure the target failure probability is not violated.

| **Variables** | **Distribution** | **Interval** | **Mean** | **CoV** |
|---|---|---|---|---|
| $b$ (mm) | Deterministic | [100,400] | — | — |
| $h$ (mm) | Deterministic | [100,400] | — | — |
| $F_a$(kN) | Deterministic | 1460 | — | — |
| $E$(MPa) | Lognormal | — | 10,000 | 0.15 |
| $L$(mm) | Lognormal | — | 3000 | 0.15 |

**Table 1.**
*Information on variables.*

| $P_f^t$ | **Proposed algorithm** | | | | | **Wang-Sheng's [47]** | | | |
|---|---|---|---|---|---|---|---|---|---|
| | $\{b,h\}$ | $f^*$ | NFE | | | $y$ | $\{b,h\}$ | $f^*$ | NFE |
| 0.1 | {206.360,197.819} | 40822 | Iter1<br>1000<br>+50<br>(L1+L2) | Iter2<br>1500<br>(L1) | | 154.65 | {209.6,209.6} | 43948 | 23100 |
| 0.01 | {206.629,197.732} | 40857 | Iter1<br>1000<br>+50<br>(L1+L2) | Iter2<br>1500<br>+100<br>(L1+L2) | Iter3<br>2000<br>L1 | 167.74 | {219.1,219.1} | 47989 | 23100 |
| 0.001 | {206.847,197.661} | 40886 | Iter1<br>1000<br>+50<br>(L1+L2) | Iter2<br>1500<br>+100<br>(L1+L2) | Iter3<br>2000<br>+100<br>(L1+L2) | 183.24 | {223.2,223.2} | 49829 | 23100 |
| | | | Iter4<br>3000<br>+100<br>(L1+L2) | Iter5<br>4000<br>L1 | | | | | |

**Table 2.**
*Results summarized for three cases with different $P_f^t$.*

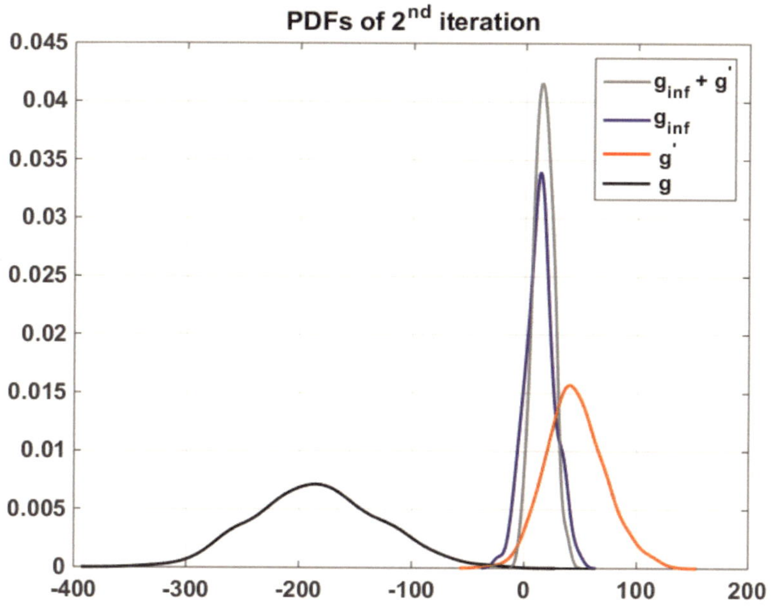


**Figure 7.**
*Pdfs of the kernel distribution fit to the random samples for second iteration* $\left(P_f^t = 0.01\right)$ *(example 1).*

Calling samples of nearby design and generating samples with a focus on failure region has led to significant improvement in the effective samples of constraint's quantile. The contribution of historical samples, L-1 sampling and L-2 sampling in the final distribution is 0.1%, 0.2% and 0.5% of the whole existing samples, respectively. **Figure 7** is provided to illustrate the contribution of each batch of samples in the failure region.

The amount of required shift for the constraint's distribution within successive iterations to reach $P_f^t = 0.01$ can be traced in **Figure 8**. It is obvious from **Figure 8** that how the algorithm is able to shift the probability constraint to the safer region with respect to the required level of reliability. In each cycle, since the failure region at the tail of the probability constraint gets narrower, the contribution of the generated samples increases, while the contribution of the historical data from nearby designs drops.

In case of setting a fixed sample size for all iterations and without recycling past information for the reliability analysis, the algorithm should produce much more samples to be able to converge to the reliable optimal point, and this will put an emphasis on the efficiency of the proposed sampling algorithm. As can be observed from **Figure 9**, without the defined adjustments in sampling levels, a sample size of 5000 is required in each iteration (a total of 25,000 for a full RBDO) to get the desired failure probability ($P_f^t = 0.001$). However, with the measurements which are consistent with a quantile-based sequential RBDO, the total number of function evaluations is as few as 11,350, that is to say, 2.2 times less computational load without sacrificing the accuracy.

## 3.2 Second RBDO problem: a highly nonlinear RBDO problem:

This problem is very well studied in various papers [48, 49], and the results of the proposed method are compared with an MCS-RBDO.

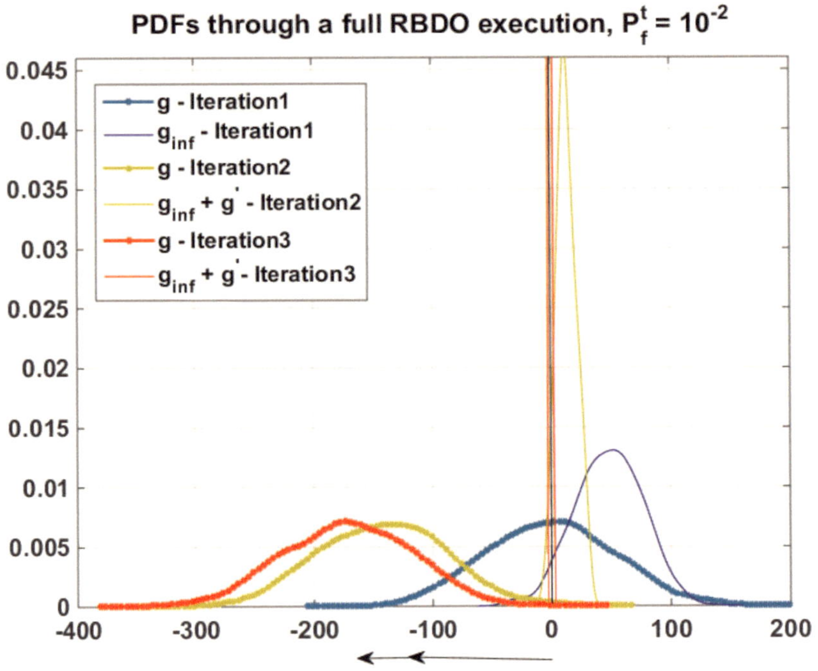


**Figure 8.**
*Pdfs of the kernel distribution fit to the random samples of all RBDO iterations for example 1 and for its second case ($P_f^t = 0.01$).*

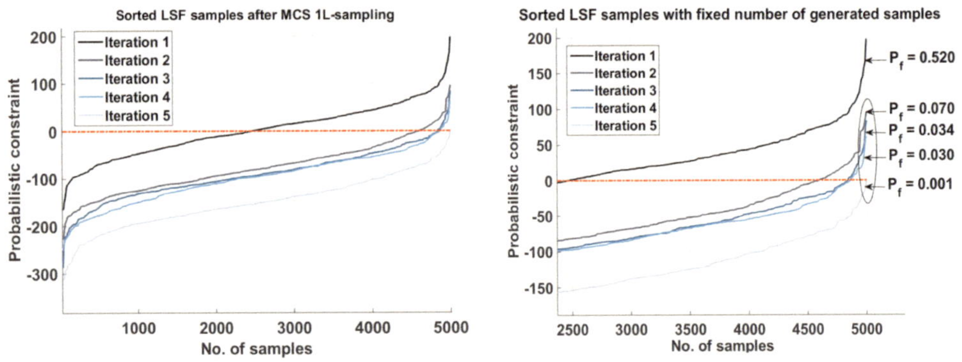


**Figure 9.**
*A great number of samples, without the proposed strategy, required for each iteration to achieve the target failure probability (here $P_f^t = 0.001$). Right panel is a zoomed view of the left panel (example 1).*

The considered mathematical design problem is formulated as

$$\textit{Minimize} \quad f(x) = -\frac{\left(\mu_{x_1} + \mu_{x_2} - 10\right)^2}{30} - \frac{\left(\mu_{x_1} + \mu_{x_2} + 10\right)^2}{120}$$

$$s.t. \quad P_f\left(g_i(x) > 0\right) \le P_{f,i}^t, \qquad i = 1, 2, 3$$

$$g_1(x) = 1 - \frac{x_1^2 x_2}{5}$$

$$g_2(x) = 1 - \frac{(x_1 + x_2 - 5)^2}{30} - \frac{(x_1 - x_2 - 12)^2}{120}$$
$$g_3(x) = 1 - \frac{80}{(x_1^2 + 8x_2 + 5)} \quad (13)$$

Which is a minimization problem including three nonlinear constraints. The input design variables $x_1$and $x_2$ are random with normal PDFs and standard deviation of 0.3 for each of them. The target failure probability is set to 2.28% for each constraint which is equivalent to 97.72% reliability. The deterministic optimal answer is taken as the initial point. The results are presented in **Table 3**.

Youn et al. in [49] have pointed out that this problem could be a good benchmark example to be tested for RBDO methods because it has a highly nonlinear and nonmonotonic performance function $g2$ as one of the active constraints. Our proposed method has shown to have good stability in guiding the constraint boundaries toward their final locations. It is apparently observed that the proposed framework requires fewer LSF calls to reach the desired level of reliability. Therefore, it can save more computational efforts than a directional MCS-based RBDO algorithm.

To give an insight into the deterministic and probabilistic constraints, **Figure 10** illustrates the contours of the design space. The initial design point and the optimal points for DDO and RBDO are shown in this figure as well. The results of the first run (reported in **Table 3**) are obtained after three iterations. The iterations progress can be seen in **Figure 11**.

The third constraint is not depicted in **Figure 11** because the algorithm has removed the third constraint that was probabilistically inactive. The remaining active constraints are shifted in each iteration until the desired level of reliability is reached. The total amounts of shifting value for these active constraints are 1.171 and 0.166, respectively.

### 3.3 Third RBDO problem: highly nonlinear limit state with random and deterministic variables

This specific benchmark problem has been selected to investigate the convergence ability of the framework [50]. This problem is characterized by two normally distributed variables as well as two deterministic design variables in a highly nonlinear

| Design vars. | $x^{LB}$ | std | $x^{UB}$ | Initial value | DDO | RIA [46] | MCS-RBDO | 2L-sampling |
|---|---|---|---|---|---|---|---|---|
| $x_1$ | 0 | 0.3 | 10 | 3.5 | 2.44 | 2.251 | 2.258 | 2.306 (2.311) |
| $x_2$ | 0 | 0.3 | 10 | 5 | 0.84 | 1.969 | 1.970 | 2.041 (2.030) |
| $f$ | — | — | — | –0.677 | –2.627 | –1.995 | –1.993 | –1.943 (–1.948) |
| $P_f(g_1)$% | — | — | — | 0 | 50 | 2.27 | 2.44 | 1.32 (1.35) |
| $P_f(g_2)$% | — | — | — | 0 | 50 | 2.28 | 2.04 | 2.02 (2.20) |
| $P_f(g_3)$% | — | — | — | 0 | 0 | 0 | 0 | 0 (0) |
| NFE | — | — | — | — | 39 | 2964 | $6 \times 10^6$ | 1400 (1600) |

**The number in parenthesis of last column denotes results of the second run.*

**Table 3.**
*Optimization results on the second RBDO problem.*

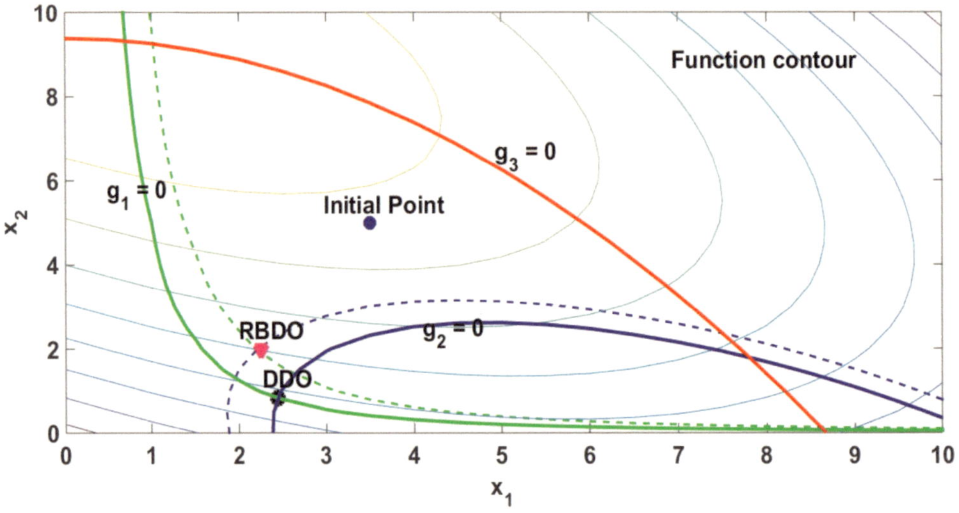


**Figure 10.**
*Results of the proposed method in the design space, the boundaries of deterministic (solid lines) and probabilistic (dotted lines) constraints (example 2).*

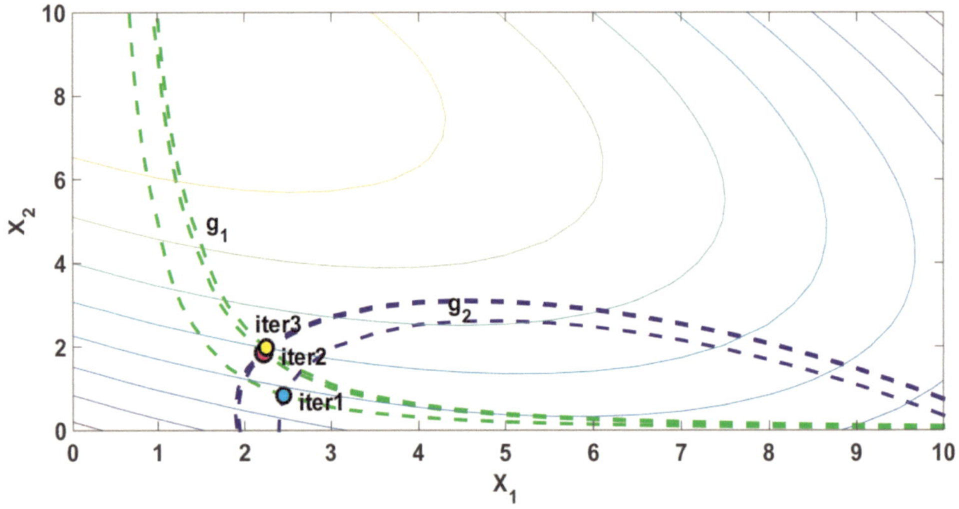


**Figure 11.**
*Active constraints in three iterations and optimal points of each iteration (example 2). The final optimal point emerges in iteration 3.*

design space. This problem is mathematically expressed as Eq. (14). The first random variable $x_1$ has mean 5 and standard deviation 1.5, whereas the second variable $x_2$ has mean and standard deviation equal to 3 and 0.9, respectively.

$$Find \quad d = [d_1, d_2]$$
$$Minimize\, f(d) = d_1^2 + d_2^2$$
$$s.t. \quad Pr(g(x,d) \leq 0)) \leq \varphi(-\beta^t)$$

$$0 \le \{d_1, d_2\} \le 15$$
$$g(x, d) = d_1 d_2 x_2 - \ln(x_1) \tag{14}$$

The optimization results of the proposed algorithm are reported in **Table 4** alongside the reported result by Shayanfar et al. in [51]. With our proposed algorithm, the achieved optimal solution is [1.35, 1.35] which is consistent with the results obtained from methods RIA, PMA, SLA and SORA reported in [50] as well as with the particle swarm optimization-wolf search algorithm (PSO-WSA) proposed by Shayanfar et al. [51] (**Table 4**) are the same and all approaches converged to the optimal point [1.35, 1.35]. Our method is robust as it shows no sensitivity to the choice of the initial point. For example, the result of **Table 4** is related to the initial point $x_0 = [4, 5]$.

Solving the DDO is dependent on the initial point, and this issue can cause problems when getting into the reliability assessment part. To reduce the dependency on the initial point, there was a need to choose an appropriate deterministic constraint handling. In order to make a balance between the effect of shift value and the error in failure probability, a cost function in the form of an exterior penalty function Eq. (8) is defined. This adjustment makes it possible for the algorithm to change the importance of the two terms adaptively, and hence, the dependency of the RBDO problem on the initial point is solved in this way.

**Figure 12** depicts a scheme of the objective function and constraint function in the allowed ranges of deterministic design variables.

| **Proposed algorithm** | | | | **Shayanfar's [51]** | | | |
|---|---|---|---|---|---|---|---|
| $\{x_1, x_2\}$ | $f^*$ | $P_f$ | NFE | $\{x_1, x_2\}$ | $f^*$ | $P_f$ | NFE |
| {1.35,1.35} | 3.67 | 0.01 | 2000 | {1.35,1.35} | 3.67 | 0.01 | 25,000 |

**Table 4.**
*Optimization results of the proposed algorithm on the third RBDO problem in comparison with PSO-WSA.*

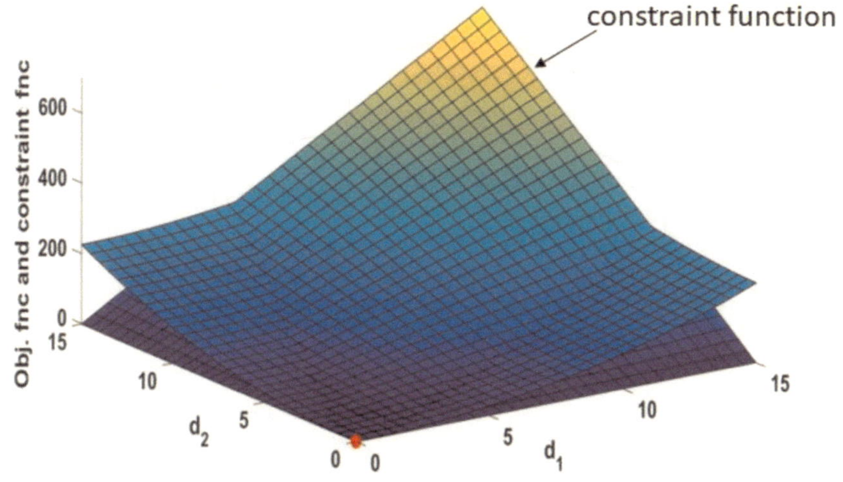


**Figure 12.**
*Design space of the third RBDO problem with the deterministic optimal point in red.*

If the constraint is properly handled in the DDO problem, then from the first iteration the probabilistic constraint will be violated. Apart from sensitivity of solving the DDO, the reliability assessment also encounters difficulties from the moving direction of the constraint. When the algorithm is running, in each iteration a specified shifting value is added to the deterministic constraint. The optimal point in each iteration moves toward the safe region, due to the structure of the problem, shifting occurs in an inappropriate direction. Consequently, instead of increasing the solution's reliability, the algorithm faces an increase the failure probability. **Figure 13** demonstrates the direction of moving constraint and optimal point for the first and second iterations. From this figure, it is clear that the optimal point has dropped into the infeasible region (**Figure 13b**.second iteration) due to the wrong direction.

This problem arises from the nonlinearity and the fact that there is no touch between the constraint function and the objective function at the points near the optimal solution. Some simple changes in the algorithm solve this issue such as replacing $g + y$ with $g - y$ in the optimization problem (**Figure 5**). Thereby, instead of reducing the failure probability, shifting values should be searched for which it leads to maximize the probability of success. In this case, necessary changes to the algorithm would be as the following:

1. $P_f < 5\% \quad \rightarrow \quad P_{success} > 95\%$

2. Statistical properties of new samples for building $g_{inf}$:

$$\text{Mean: } \mu_{sort\ x\ (x_{mpp}\ :x_{end})} \quad \rightarrow \quad \mu_{sort\ x\ (x_1:x_{mpp})}$$

$$\text{Std: } \sigma_{sort\ x\ (x_{mpp}:x_{end})} \quad \rightarrow \quad \sigma_{sort\ x\ (x_1:x_{mpp})}$$

3. $g_f > y \quad \rightarrow \quad g_f < -y$

4. $P_f = \frac{1}{N}\sum_{i=1}^{N} \mathbb{I}_{inf}(d, X)$

$$\mathbb{I}_{inf}(d, X) = \begin{cases} 1, & x \in \inf \le -y, \\ 0, & \text{else.} \end{cases}$$

5. $P_f^t = 5\% \quad \rightarrow \quad P_{success}^t = 95\%$

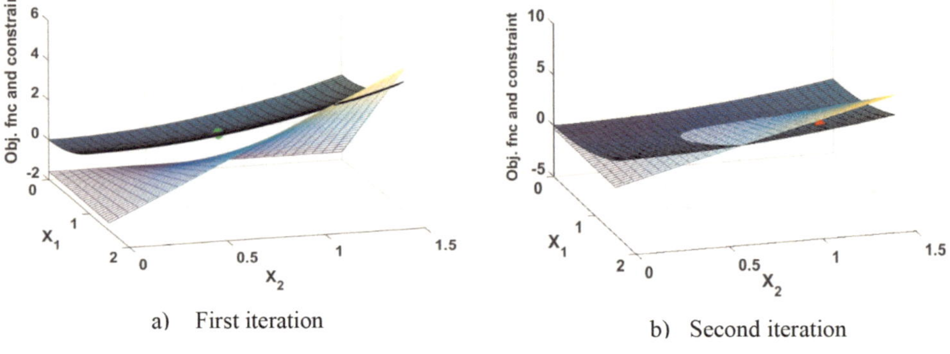


**Figure 13.**
*Design space of the third RBDO problem for two consecutive iterations.*

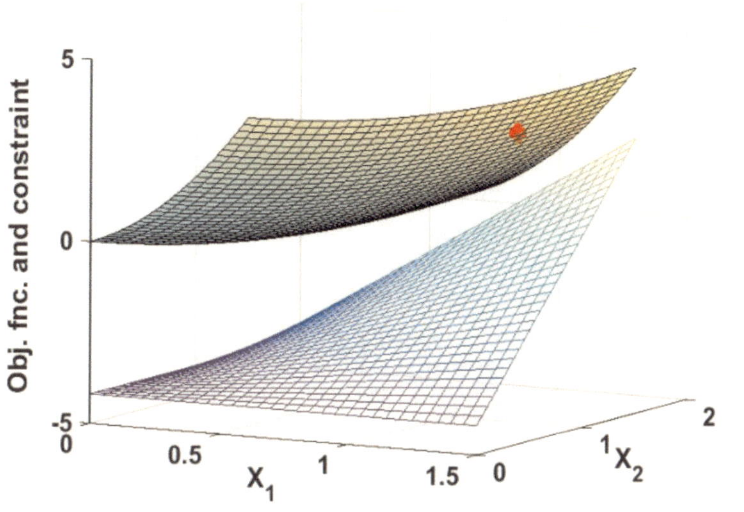


**Figure 14.**
*Design space of the third RBDO problem after making changes to reverse the direction of shifting constraint (last iteration).*

By making the above changes in the algorithm, the direction of shifting constraints will be reversed. See **Figure 14**.

In **Figure 15**, as expected, the algorithm has sharply reached a high probability of failure at the very beginning when the reliability is still very low. Therefore, according to Eq. (5), the shift value term has outweighed the failure probability error. As the algorithm approaches the last iterations and hence the narrower the failure region, the algorithm performs the constraint's shifting with smaller sizes until it minimizes the

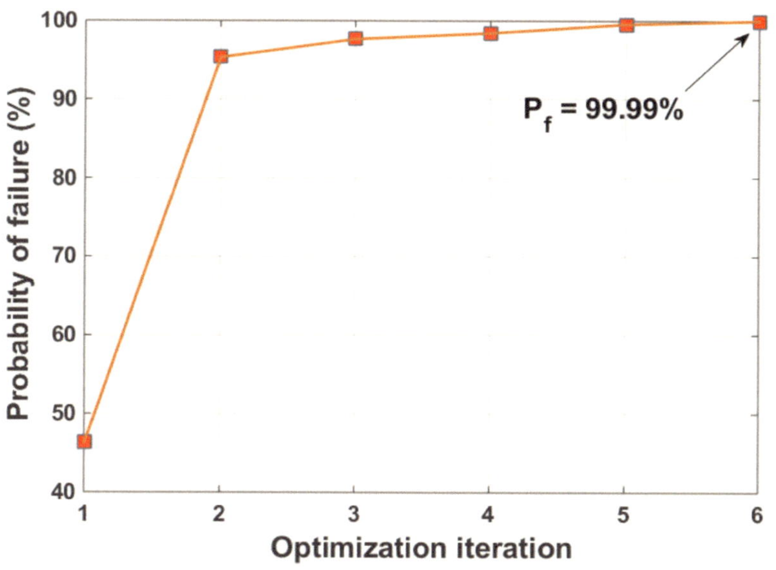


**Figure 15.**
*Failure probability in each optimization iteration for the third RBDO problem.*

failure probability error. Another interesting point is that as opposed to many other sampling-based RBDO [30, 44], the proposed algorithm ensures a more reliable point in each iteration compared to previous iterations. According to **Figure 15**, the algorithm has converged from high errors in failure probability to lower errors as the algorithm progresses.

Finally, from a top-down view, the proposed method has several advantages: First, due to adopting a sequential RBDO structure, only constraints which become active in each iteration will undergo reliability analysis which is not the case of double-loop methods. This causes a significant computational savings. Second, thanks to adopting sampling and quantile strategy in the estimation of failure probability, there is no need to struggle with searching a global MPP, particularly in highly nonlinear problems, as opposed to the existing sequential RBDO. Third, due to the adaptive step size introduced in the shifting value search strategy, we can make sure that a more reliable optimal point will be obtained in each iteration without any concern about the overdesign issue.

**Table 5** is a general comparison between the proposed method and the existing methods which clearly show the applicability of the algorithm.

| Feature | Reviewed methods in literature | Proposed model |
|---|---|---|
| Classified in which category?<br>• Double-loop strategies<br>• Decoupled strategies<br>• Sampling-based strategies | Double loop: [6, 7]<br>Decoupled: [9–13, 42, 43]<br>Sampling: [20–44, 46–48, 51] | More into:<br>• Sampling and<br>• Single-loop methods |
| The way of calculating failure probabilities?<br>• Simulation-based<br>• Numerical integration<br>• Analytical methods | Simulation-based: [20–44, 46–48, 51]<br>Numerical: [17, 18]<br>Analytical: [15, 16] | • Simulation-based |
| Need to design a proposal distribution for sampling? | IS: [20, 21, 24, 27, 30, 41, 44],<br>SS: [22, 25, 26, 28, 29] | • No<br>(In doing so, it is beneficial for high dimensional problems as well as nonlinear LSF especially for small failure probabilities and also non-Gaussian random variables.) |
| Construct surrogate models to estimate required objective functions? | Data-driven methods: [31–38, 42] | • No<br>(In being so, there is no need for manipulating the approximate model close to the failure region in order to avoid losing desired accuracy.) |
| Require calculating the constraints/functions gradient? | Analytical methods [15, 16] | • No<br>(Beneficial in case of highly nonlinear and non-differentiable functions.) |
| Information reuse? | [30, 44] | • Yes<br>(But as opposed to the reviewed literature, information reuse is employed in sequential RBDO with its inherent benefits.) |

**Table 5.**
*General features of the proposed method in summary.*

## 4. Conclusion

Inspired by SORA's method, a novel sequential RBDO method is proposed which integrates the shifting constraint's strategy with quantile-based probability of failure estimation instead of MPP. Searching for global MPP is problematic when being performed in the random design space. Hence, in the proposed method, the RBDO structure is disintegrated in the probability space instead. This quantile-based algorithm uses a combination of different techniques to ensure a better result is obtained in each iteration so that wherever the problem is paused, it ensures us that the current design is more reliable than the previously generated ones. Once the deterministic optimization problem is solved, the failure probability estimation process in the probability space begins which is based on the quantile calculation with sampling. To enhance the computational efficiency in sampling, in the first level, the samples are generated adaptively. Then, the second-level sampling is done focusing on the failure region with the knowledge obtained from first-level sampling. Furthermore, the samples of previous iterations are reused to build a mixture distribution for which most of the samples are distributed in the failure region to provide an accurate estimate of the quantile. After the quantile is calculated, with an efficient adaptive step size for the optimal shifting value, the algorithm is able to quickly converge to the target failure probability. We explored the efficiency of the proposed method through three benchmark problems. The results show that the proposed method will decrease the computational cost to less than half of what is reported about other existing methods in very small target probabilities ($P_f^t = 0.001$) and down to 9% in mild target probabilities ($P_f^t = 0.1$ and $P_f^t = 0.01$)). It is also able to provide promising results both for highly nonlinear problems and problems with mixture of normal–nonnormal variables as well as random-deterministic variables. As an extension to this work, the same framework will be combined with the IS concept to reach a potentially more efficient algorithm for smaller failure probabilities, and we will explore it in the future works.

## Acknowledgements

This work was conducted with the support of the Science Foundation Ireland Centre for Research Training in Artificial Intelligence under Grant No. 18/CRT/6223.

## Disclosure statement

No potential conflict of interest was reported by the author(s).

## Author details

Shima Rahmani[1], Elyas Fadakar[2] and Masoud Ebrahimi[3*]

1 School of Computer Science and Statistics, Trinity College Dublin, Dublin, Ireland

2 Electronics and Information Engineering, Beihang University (BUAA), Beijing, China

3 Faculty of Mechanical Engineering, Tarbiat Modares University, Tehran, Iran

*Address all correspondence to: ebrahimikm@modares.ac.ir

## References

[1] Beck AT, Gomes WJS, Bazán FAV. On the robustness of structural risk optimization with respect to epistemic uncertainties. International Journal for Uncertainty Quantification. 2012;**2**(1):1-20

[2] Lopez RH, Beck AT. Reliability-based design optimization strategies based on FORM: a review. Journal of the Brazilian Society of Mechanical Sciences and Engineering. 2012;**34**:506-514

[3] Tu J, Choi KK, Park YH. A new study on reliability-based design optimization. Journal of Mechanical Design. Dec 1999; **121**(4):557-564 (8 pages)

[4] Du X, Chen W. Sequential optimization and reliability assessment method for efficient probabilistic design. Journal of Mechanical Design. 2004; **126**(2):225-233

[5] Agarwal H et al. An inverse-measure-based unilevel architecture for reliability-based design optimization. Structural and Multidisciplinary Optimization. 2007;**33**(3):217-227

[6] Torii AJ, Lopez RH, Miguel LFF. On the accuracy of reliability index based approaches and its application to RBDO problems. In: 22nd International Congress of Mechanical Engineering (COBEM 2013), Brazil. 2013

[7] Kuschel N, Rackwitz R. Two basic problems in reliability-based structural optimization. Mathematical Methods of Operations Research. 1997;**46**(3):309-333

[8] Huang ZL et al. An incremental shifting vector approach for reliability-based design optimization. Structural and Multidisciplinary Optimization. 2016;**53**(3):523-543

[9] He Q et al. Reliability and multidisciplinary design optimization for turbine blade based on single-loop method. Tuijin Jishu/Journal of Propulsion Technology. 2011;**32**(5)

[10] Liang J, Mourelatos ZP, Tu J. A single-loop method for reliability-based design optimization. In: International design engineering technical conferences and computers and information in engineering conference. 2004; 46946

[11] Jiang C et al. An adaptive hybrid single-loop method for reliability-based design optimization using iterative control strategy. Structural and Multidisciplinary Optimization. 2017;**56**: 1271-1286

[12] Meng Z et al. Convergence control of single loop approach for reliability-based design optimization. Structural and Multidisciplinary Optimization. 2018;**57**: 1079-1091

[13] Keshtegar B, Hao P. Enhanced single-loop method for efficient reliability-based design optimization with complex constraints. Structural and Multidisciplinary Optimization. 2018; **57**(4):1731-1747

[14] Yao W et al. Review of uncertainty-based multidisciplinary design optimization methods for aerospace vehicles. Progress in Aerospace Sciences. 2011;**47**(6):450-479

[15] Madsen H, Krenk S, Lind N. Methods of Structural Safety. Englewood Cliffs: Prentice-Hall; 1986

[16] Melchers RE, Beck AT. Structural Reliability Analysis and Prediction. UK: John Wiley & Sons; 2018

[17] Lee SH, Kwak BM. Response surface augmented moment method for efficient

reliability analysis. Structural Safety. 2006;**28**:261-272

[18] Rahman S, Xu H. A univariate dimension-reduction method for multi-dimensional integration in stochastic mechanics. Probabilitic Engineering Mechanics. 2004;**19**:393-408

[19] Youn BD, Zhimin X. Reliability-based robust design optimization using the eigenvector dimension reduction (EDR) method. Structural and Multidisciplinary Optimization. 2009;**37**: 475-492

[20] Au SK, Beck JL. A new adaptive importance sampling scheme for reliability calculations. Structural Safety. 1999;**21**(2):135-158

[21] Kanj R, Joshi R, Nassif S. Mixture importance sampling and its application to the analysis of SRAM designs in the presence of rare failure events. In: Proc. Design Autom. Conf., ACM, 2006

[22] Au S-K, Beck JL. Estimation of small failure probabilities in high dimensions by subset simulation. Probabilistic Engineering Mechanics. 2001;**16**(4): 263-277

[23] Elsheikh A, Oladyshkin S, Nowak W, Christie M. Estimating the probability of CO2 leakage using rare event simulation. In: ECMOR XIV-14th European conference on the mathematics of oil recovery. 2014

[24] Shayanfar MA et al. An adaptive directional importance sampling method for structural reliability analysis. Structural Safety. 2018;**70**:14-20

[25] Li H-S, Cao Z-J. Matlab codes of Subset Simulation for reliability analysis and structural optimization. Structural and Multidisciplinary Optimization. 2016;**54**(2):391-410

[26] Li HS, Ma YZ. Discrete optimum design for truss structures by subset simulation algorithm. Journal of Aerospace Engineering. 2015;**28**(4): 04014091

[27] Yu T, Lu L, Li J. A weight-bounded importance sampling method for variance reduction. International Journal for Uncertainty Quantification. 2019;**9**(3)

[28] Li L, Bect J, Vazquez E. Bayesian subset simulation: A kriging-based subset simulation algorithm for the estimation of small probabilities of failure. In: 11th International Probabilistic Assessment and Management Conference (PSAM11) and The Annual European Safety and Reliability Conference (ESREL 2012), Helsinki: Finland. 2012. arXiv:1207.1963

[29] Au SK, Wang Y. Engineering Risk Assessment with Subset Simulation. Singapore: John Wiley & Sons; 2014

[30] Chaudhuri A, Kramer B, Willcox KE. Information reuse for importance sampling in reliability-based design optimization. Reliability Engineering & System Safety. 2020;**201**: 106853

[31] Motamed M. A multi-fidelity neural network surrogate sampling method for uncertainty quantification. International Journal for Uncertainty Quantification. 2020;**10**(4):315-332

[32] Zhao Y, Lu W, Xiao C. A kriging surrogate model coupled in simulation-optimization approach for identifying release history of groundwater sources. Journal of Contaminant Hydrology. 2016;**185-186**:51-60

[33] Mi X, Zhang J, Gao L. A system active learning Kriging method for system reliability-based design

optimization with a multiple response model. Reliability Engineering & System Safety. 2020;**199**:106935

[34] Jiang C et al. Global and local Kriging limit state approximation for time-dependent reliability-based design optimization through wrong-classification probability. Reliability Engineering & System Safety. 2021;**208**: 107431

[35] Ciriello V, Di Federico V, Riva M, Cadini F, De Sanctis J, Zio E, et al. Polynomial chaos expansion for global sensitivity analysis applied to a model of radionuclide migration in a randomly heterogeneous aquifer. Stochastic Environmental Research and Risk Assessment. 2013;**27**(4):945-954

[36] Gao T, Li J. A cross-entropy method accelerated derivative-free RBDO algorithm. International Journal for Uncertainty Quantification. 2016;**6**(6): 487-500

[37] Zhao W, Qiu Z. An efficient response surface method and its application to structural reliability and reliability-based optimization. Finite Elements in Analysis and Design. 2013; **67**:34-42

[38] Yang I-T, Hsieh YH. Reliability-based design optimization with cooperation between support vector machine and particle swarm optimization. Engineering with Computers. 2013;**29**(2):151-163

[39] Smith RC. Uncertainty Auantification: Theory, Implementation, and Applications. Philadelphia, PA: SIAM; 2013

[40] Alyanak E, Grandhi R, Bae H-R. Gradient projection for reliability-based design optimization using evidence theory. Engineering Optimization. 2008; **40**(10):923-935

[41] Wang Z, Song J. Cross-entropy-based adaptive importance sampling using von Mises-Fisher mixture for high dimensional reliability analysis. Structural Safety. 2016;**59**:42-52

[42] Li G, Yang H, Zhao G. A new efficient decoupled reliability-based design optimization method with quantiles. Structural and Multidisciplinary Optimization. 2020; **61**(2):635-647

[43] Yi P, Xie D, Zhu Z. Reliability-based design optimization using step length adjustment algorithm and sequential optimization and reliability assessment method. International Journal of Computational Methods. 2019;**16**(07): 1850109

[44] Beaurepaire P et al. Reliability-based optimization using bridge importance sampling. Probabilistic Engineering Mechanics. 2013;**34**:48-57

[45] Fischer H. A History of the Central Limit Theorem: from Classical to Modern Probability Theory. Springer Science & Business Media; 2010

[46] Chen Z, Li X, Chen G, Gao L, Qiu H, Wang S. A probabilistic feasible region approach for reliability-based design optimization. Structural and Multidisciplinary Optimization. 2017

[47] Liu W-S, Cheung SH. Reliability based design optimization with approximate failure probability function in partitioned design space. Reliability Engineering & System Safety. 2017;**167**: 602-611

[48] Kim D-W et al. A single-loop strategy for efficient reliability-based

electromagnetic design optimization. IEEE Transactions on Magnetics. 2015;**51**(3)

[49] Youn BD, Choi KK, Du L. Enriched performance measure approach for reliability-based design optimization. AIAA Journal. 2005;**43**(4): 874-884

[50] Aoues Y, Chateauneuf A. Benchmark study of numerical methods for reliability-based design optimization. Structural and Multidisciplinary Optimization. 2010;**41**(2):277-294

[51] Shayanfar MA, Barkhordari MA, Roudak MA. An efficient reliability algorithm for locating design point using the combination of importance sampling concepts and response surface method. Communications in Nonlinear Science and Numerical Simulation. 2017;**47**: 223-237

Section 3

# Artificial Intelligence Applications

Chapter 3

# Deep Neural Networks for Unsupervised Robotics in Building Constructions: A Priority Area of the Fourth Industrial Revolution (4IR)

*Nicholas Eze, Ekene Ozioko and Johnpaul Nwigwe*

## Abstract

Many effective quality systems to maintain the robots' autonomous task expansion process in construction industries for various applications over the years have yet to be well established. This study, therefore, presents a simple deep/neural network algorithm to diverse robotics tasks on building construction—bricklaying, grasping, cutting materials, and aerial robot obstacle avoidance and highlight the strengths of these algorithms in real-world robotics applications in building sites. Our findings revealed that the amount of tasks robots encountered in real-world environments is extremely challenging for existing robotic control algorithms to handle. Also, our algorithm when evaluated against other conventional learning algorithms can be a more powerful tool with the capacity to learn features directly from data, making it an excellent choice for such robotics applications in building construction. In other words, our algorithm can teach robots the ability to "work," "think," "know," and "understand" their surroundings. It can also improve customer satisfaction, speed up the building process, and improve the productivity of building development teams. This chapter, however, contributes to classifications of autonomous robotics application development in construction literature. Although the problem addressed in this chapter is based on building construction, the algorithms presented are designed to be generalizable to related tasks.

**Keywords:** automation, building construction, construction robots, deep learning industrialization, machine learning, neural networks, robotics

## 1. Introduction

Robotics has influenced nearly every modern construction industry by improving efficiency, safety, and cost [1]. Robotics offers process automation and reliability thanks to sensor technologies [2]. However, while most industries have embraced this new technology from the moment it is released, many construction industries have historically been slow to onboard automated solutions [3]. The reason could be linked to product features and complexity (project size, lifetime and uniqueness, versatile construction environments). According to Ref. [4], several other phenomena could add to the characteristics of these construction industries that often contribute to the complexity of projects. This phenomenon includes client needs that are sometimes imprecise and changing, causing significant change in costs; little overall learning because of few repetitions; high risks due to novelty; technical, climatic, and even societal uncertainties; coordination and complex decision-making processes between the teams involved; and changing conditions of realization. The weak capital budgets in R&D and the reluctance of strategies related to construction automation are other important factors [5].

These phenomena have remained unresolved for a variety of reasons. First, building nowadays are complex entities, and construction entails many different trades coming together to work in perfect sync with each other [6–9]. A replica of human-like dexterity, intelligence, and situational awareness, developed over hundreds of millions of years, is needed to break even [10]. Secondly, building construction project sites is often chaotic, disorganized spaces with materials, tools, debris, and wires spread about. Many areas of the sites have unpaved soft soil, into which a builder may sink if he steps off the beaten path. There are some environmental factors such as dust, rain, ice, and storms. There are humans walking around, etc. All these have not improved for so long because construction industries are the least digitized sector [3, 5].

Nevertheless, several ongoing initiatives suggest a gradual change in practices in this industry for rapid industrialization, which is enabled thanks to robotics and automation. According to Ref. [11], rapid industrialization is always based on quantity and quality that includes prefabrication, mechanization, automation, robotics, and finally reproduction. This gradual change toward rapid industrialization is driven by a concern to change the narrative and to be in line with the innovations observed at the international level that will respond to important building construction challenges: workforce, competitiveness, sustainable development, etc. In view of this, digital transformation with robotics technologies is one of the preferred avenues to improve the sector's overall performance over the long term [12]. This digital transformation is expressed in different ways. One of the strong currents is based on the so-called concept of the Fourth Industrial Revolution (4IR), based on unsupervised robotics utilization in building constructions.

### 1.1 Rationale and research gap

While the robotics use in building construction will only continue to grow from traditional design through final inspection and maintenance, the full benefits of construction robotics have yet to be realized. For example, as robotics begins to move from the lab to the real world, robots face many new challenges. A building construction assistant robot, for instance, must perform many complex tasks such as bricklaying, foaming, sorting, operating appliances, picking up, and cleaning materials in the site. It must also handle the huge variety of objects, materials, and the likes associated with these tasks such as picking up different objects, some of which it may never have seen before. For all of these problems, there exists only an abstract relationship between the robot's visible inputs and the task at hand.

Traditionally, a roboticist, or team thereof, would hand-design these robots for each task they want the robots to perform. Even for tasks which human users can perform intuitively, such as bricklaying, grasping, cutting objects, detecting, and avoiding various forms of obstacles in building construction sites, and these robots can be very difficult to design because developers were not able to easily translate these abstract intuitions between the robot's visible inputs and the task at hand into code. This makes it extremely challenging to scale these approaches up to the huge amount of works and obstacles that building construction robots must deal with in the real-world construction.

Secondly, most robotics that engaged in construction activities primarily rely on a system performance metric that is dependent on a metric connected to the given human-defined tasks, or task-dependent metric. This also implies a conventional paradigm of same manual development, where human designers were in charge of planning and coding for particular tasks-based activities that the robot would perform and how it would perform them [13–15]. This has some drastic limitations on the construction industry. For example, it is impossible for a bricklaying robot to perform delicate tasks such as installing electrical cables, or even to be able to detect and avoid various types of obstacles on the building construction sites. This limitation caused during design and coding thus prevented these robots from independently "thinking," "knowing," or "understanding" the multiple building operations thus generating questions about what developers had overlooked that made them to fail to get the anticipated outcome of operational autonomy.

Several researchers have criticized this conventional architecture for its lack of computational data [14–19] though went on to propose several failed models, which would have addressed this research gap. For example, in their research, several researchers [14–16] provided the first implementations of this autonomous curiosity, but they were unable to integrate their concept within the issues with construction of robo-mason by demonstrating how the robot's work patterns could emerge without the assistance of a human.

However, in order to establish this brain model that will give the robot significant cognitive developing abilities unique to humans, this study plans on modeling the brain at a level above the neural level, or what would normally be thought of as the unsupervised or unmanned level using neural deep learning algorithms that will allow the robots to learn independently from some training data. Our understanding of brain abstraction is sufficient to program a system that exhibits similar properties and connections to the human brain without having to model its detailed local wiring. Quite clearly, we will model this machine learning algorithmic concept based on neural networks and highly-parameterized models, which will use multiple layers of representation to transform data from a task-specific representation to an autonomous task. While our algorithm is not designed to solve central problems in artificial intelligence (AI), such as speech or object recognition, its development was motivated by the need to improve the performance of robots in the building construction industry. Traditional algorithms often struggle to generalize well on AI tasks specific to this domain, which prompted us to explore the potential of deep learning.

This chapter presents a distinct application of deep neural learning algorithms that can enable a building construction robotics to learn from some training data and to perform highly cognitive artificial intelligence operations. Here, rather than forcing the engineer to hand-code an entire end-to-end construction robotic system, our machine learning algorithm will allow portions of the system to be *learned* from some training data. This approach will allow us to model concepts, which might be difficult or impossible to properly hand-model. It will also allow for adaptable models—as long as the form of the model is general, meaning it can be adapted to more or different cases simply by providing training data for these new cases.

However, by using unsupervised feature learning algorithms, deep learning approaches are able to pre-initialize these networks with useful building construction features, thus avoiding the overfitting problems commonly seen when neural networks are trained without this initialization. This machine learning algorithm can therefore work very well on a wide variety of construction projects.

Nevertheless, we begin with a general description of deep learning algorithms for unsupervised feature learning as well as their strengths and particular advantages as learning algorithms for robotics applications on building constructions. Finally, we present a simple deep/neural network algorithm to diverse robotics tasks on building construction—bricklaying, grasping, cutting materials, and aerial robot obstacle avoidance—highlighting the strengths of these algorithms in real-world robotics applications in building sites. It is our hope that these algorithms will demonstrate a more appropriate computation model where, for instance, robots' artificial intelligence and ability to detect obstacles and carry out multiple construction tasks unsupervised are no longer isolated from the subjective experience of the body.

## 2. Literature review

### 2.1 Deep learning/neural network for unsupervised robotics technologies

Deep learning is a machine learning approach based on modeling adaptation of biological neural systems. It can also be defined as a computing system made up of a number of simple, highly interconnected processing elements, which process information by their dynamic state response to external inputs [20]. Information processing is carried out through connectionist approach to computation and this amount of information needs a complex abstraction as data representations through a hierarchical learning process [21]. The term hierarchical learning here is referred to neural networks [22]. An example of such a neural network is shown in **Figure 1**.

Here, our algorithm could be implemented to train the network in an unsupervised manner. This is because the multi-layer (with many hidden layers) neural network is being used as shown in **Figure 2** in which each layer takes input from the previous layer, processes it, and outputs it to the next layer, in a daisy-chain fashion.

With this, our deep learning can be used to generate high level of abstraction for the building construction robots. So for complex abstractions of data representations through a hierarchical learning process, our deep learning model can produce results faster than standard machine learning [23]. And also our proposed deep learning model will be integrated with building construction features that are important by itself to be learned, instead of requiring to be manually selecting the pertinent features. However, this unsupervised learning model does not require the presence of a teacher. The desired output is not presented to the network. The system learns on its own by adapting the structural features in the input patterns. The general role of our unsupervised learning model is shown in **Figure 3**.

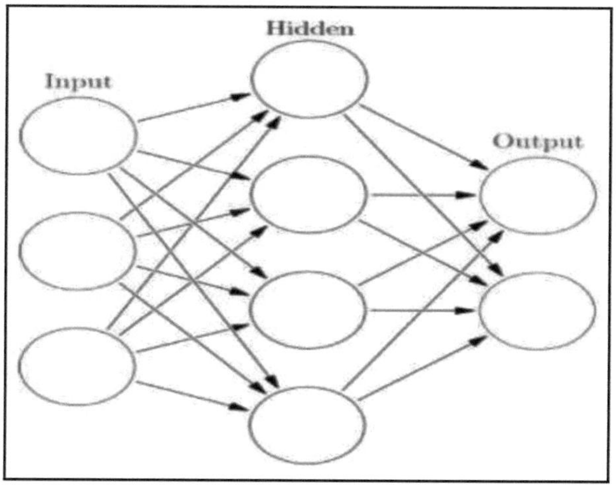


**Figure 1.**
*Multi-layer artificial neural network.*

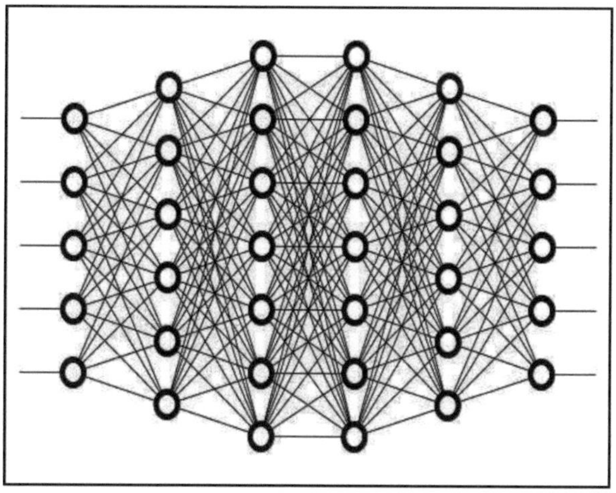

**Figure 2.**
*Deep neural network for deep learning.*

## 2.2 Robotics application and its capacity on building construction

Robotics applications in construction are a large field of study due to the multidisciplinary trades that constitute the act of constructing. For example, bridges are a type of construction subjected to various robotic automations. Oh et al. [9] developed a robot for bridge inspection and Lorenc et al. [24] for maintenance. Bridges can be difficult to access, and there is a high demand for robots to perform diagnostics and repairs. However, we have gathered information from various resources especially from relevant literatures during a span of 10 years, i.e., 2013–2023 pertaining to more forms of robotics' applications and its capacity in building construction.

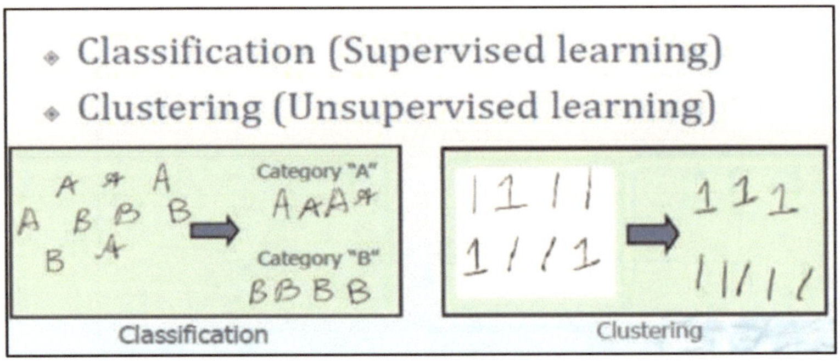


**Figure 3.**
*Classification and clustering.*

### *2.2.1 Drones/UAV technologies application in building construction*

Few numbers of references are reported in the literature regarding use of drones/UAVs technologies in construction by many authors [3, 25–27]. The reports show that the use of drones highlights the essential role of humans in robotics, as drones are being utilized for a range of tasks from painting to identifying safety concerns on work sites. See **Figure 4**. These robotic manipulators are appropriate for the different spraying activities involved in construction work. For example, high-rise building coating, fireproofing application, and shot blasting can all be done without the need for a human operator to be physically present. With the aid of drone robots, a variety

**Figure 4.**
*Drone in construction work site.*

of applications, including cleaning and sandblasting, may be completed effectively. Drones that paint walls in tall buildings are among the robotic technologies. It comes equipped with automated spraying equipment and shot blasting capabilities for surface preparation.

Also, drone data has been integrated with data captured by ground-based robots, known as Autonomous Mobile Robots (AMRs), to provide a real-time view of a work site [28]. This data can then be integrated into virtual reality, allowing project managers to see the construction site without having to leave the office.

Drone data becomes necessary because traditional inspection is labor-intensive. Due to the harsh workplace and lack of trained personnel, site inspection can be completely automated using feedback from auto survey tools. These tools have a video and electronic display that transmits data to the microcomputer using fiber optic cable and employs a laser to configure the shield machine. This drone technology has been briefly applied in Architecture, Engineering, and Construction (AEC) domain and provided guidance for UAV operation and implementation in the construction industry [29]. Scaled Robotics are forms of drone which are developed as Autonomous Mobile Robots (AMRs) that are controlled *via* mobile devices where information is collected by robots using laser scanners, focusing on identifying obstacles, mistakes, or errors in an effort to reduce rework [30]. This is done with the aid of some digital photogrammetric systems that are combined with robots for demolition and site cleanup to manage large-scale earthwork projects. They are also remote-controlled vehicles featuring microwave communications, radar beacons for position, and radar sensors for avoiding obstructions and identify errors. They are, therefore, capable of controlling up to 100 units simultaneously and can identify impediments at a distance of up to 2 kilometers.

Again [31, 32], extensively conducted a systematic literature review on the current topic pertaining to implementation of Unmanned Aerial System (UAS) in the construction industry covering the most relevant job, cases, and areas of application. Their report revealed that the recent developments in UAS regulations have played a significant role in their popularity and wide deployment in various stages of the construction lifecycle. A comprehensive study conducted in the United States to identify the practical construction UAS application areas, their adopted technologies, as well as the benefits and barriers encountered during their implementation [33], and further revealed that drone robotics technologies today help many aspects of the building construction industries, including knowledge-based design and control [34].

However, a critical literature review has also been carried out on the relevant existing studies toward this immersive and digital technology [6]. The authors analyzed the literature using meta-synthesis technique to evaluate and integrate the findings in a single context. Their review shows that this class of construction drone robots for various purposes focuses mostly on tasks in hazardous areas where people cannot perform them. These domains include deep sea oil prospecting, nuclear power plants, and decommissioning issues. However, counter reports have revealed that these immersive and digital technologies have not been significantly utilized in the construction sector and by extension of other sectors of the economies in developing countries due to the high level of investment required [35–38]. It therefore makes sense to utilize these technologies that can function without administrative or motivational problems in hazardous conditions, bad weather, and at night to reduce the growing number of accidents in hazardous building tasks [39].

### *2.2.2 Bricklaying robots*

Bricklaying work is one of the most arduous jobs in construction [40, 41] since it includes a mason standing, kneeling, and lifting. In addition, the mason works almost exclusively outside and undergoes the weather conditions (rain, wind, heat, humidity). The mason sometimes works in height scaffolding or in trenched soils, which may put his life in jeopardy. However, in the last two decades, some research projects focused on the development of a bricklaying robot [42]. Bricklaying work follows predefined steps and thus is favorable for automation. However, the process cannot be fully automated and requires the supervision of a worker nearby to adjust/control the robot. Tan et al. [43] stressed the importance of the environment when designing a robot. They support the idea that robot level of autonomy should be in line with the environment (actively/passively/not assisted environment). For that, the authors proposed a framework to help categorize the robot/environment interaction.

Recent advancements in masonry work automation technology include Australian Hadrian X and Construction Robotics' SAM100 also known as semi-autonomous mason are robotic bricklaying machines [44, 45]. Hadrian X uses an intelligent control system alongside Computer Aided Drawing (CAD) to function and is capable of building a standardized home every two days on average. The robot is capable of laying the bricks with a high accuracy thanks to a laser guidance system. It is also able to work on almost any block size. The advantage of such a design is the flexibility in mobility: The robot can work under difficult circumstances linked to the environment. That is to say, by deploying Hadrian on a construction project, one can benefit from faster masonry work, least material wastage, and overall cost-efficacy. Hadrian X closely resembles a truck crane (**Figure 5**).

SAM100 (Semi-Automated Mason) on the other hand is designed to work collaboratively with masons to increase productivity by 3 to 5 times while reducing lifting by 80% and so on [45]. The robot has successfully passed the prototyping phase and is

**Figure 5.**
*Bricklaying robot "HADRIAN X." source: Pivac and Pivac [44].*

**Figure 6.**
*SAM100 robot onsite screenshot. Source: Podkaminer and Peters [45].*

now commercially available. **Figure 6** shows the utilization of SAM100 onsite. This robot is by far the most complete masonry robot realized until now. It can lay bricks with precision and includes the binder in the process of laying as well. SAM100 is capable of laying 800–1.200 bricks a day. The robot performs in a straight line with a limited height capacity. SAM100 costs around 500.000$ (442.030 €).

### *2.2.3 Specific design building construction robots*

Several other research projects have focused on a specific design problematic of the automation process. For example, SAM100 design is based on an articulated arm as found in previous research projects [12, 46]. The "HADRIAN X" is based on a variant of the articulated arm supported by a truck-crane robot. FUNAC robots and COGIRO robots with long arms were also not left out in the discussions. COGIRO robot is used as a precise tele-operated crane to position prefabricated roof elements, while the FUNAC robots does monotonous, risky, repetitive construction works, handling payloads up to 2300 KG due to the strength of its axis, see **Figure 7**.

These robots promise to reduce operating costs and waste, as well as provide safer work environments and improve productivity. However, while deploying robots like the FUNAC in a building construction project, the number of axes it has must be a crucial factor to be considered [39]. Many people might not be aware of the significance of the robots' axes and how they regulate a robot's range of motion and strength of work. Each axis of a robot stands for a degree of freedom, or to put it another way, an independent motion that enables a construction robot to become more functional. In other words, the more degrees of freedom and higher usefulness a robot has, the more axes it has. For instance, six-axis FUNAC robots are perfect for repetitive, tiresome, and even dangerous construction tasks that were previously solely performed by people because designers mirror human arm movements and offer the same degrees of freedom as human arms [47].

**Figure 7.**
*FUNAC robots with long arms placing bricks and other materials.*

FUNAC robots can therefore do whatever their human counterparts can do in building construction, without becoming fatigued or risking their safety. They have better access to even challenging vocations due to their wider range of motion and numerous degrees of freedom, thus they carry out several repetitive operations. It is perfect for the majority of applications found on construction sites, grasping, and handling payloads up to 2300 KG and reaching beyond 3.5 meters. In **Figure 8** below, FUNAC robot models are displayed.

These six-axis robots offer more advantages than other types since they have a wider range of motion and can move in more directions than just the x, y, and z planes owing to their various degrees of freedom. For example, a three-axis robot can only move in three planes (x, y, and z) because it lacks the other three axes. Robots with four or five axes may move in all three planes and can even perform extra actions like rotating or elevating the mixers.

However, due to the minimal amount of movement required to remove something from a conveyor and place it on a pallet, four-axis robots are frequently utilized in

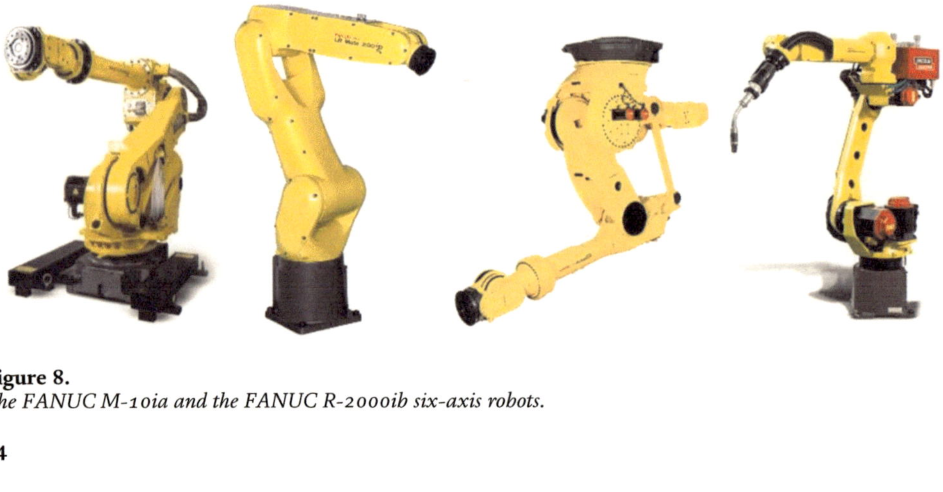

**Figure 8.**
*The FANUC M-10ia and the FANUC R-2000ib six-axis robots.*

palletizing applications. Although practically any palletizing operation may be completed using the four-axis FANUC M-410ib/160 without the use of the 2 extra axes, it has been previously shown [39] that a robot's range of motion increases with each axis it possesses.

In sum, we have conducted this extensive literature work to raise the consciousness of our readers on what the literature said regarding the applications and strengths of robotics in building construction. Our conclusion is that when applying robotics in building construction, engineers have a lesser work envelope because these machines are more versatile and have a wider range of work alternatives.

### 2.3 Unsupervised robots in building construction: Requirements and important qualities

#### *2.3.1 Robot requirements for building construction tasks*

- It should be effective in preventing human operators from dying in dangerous circumstances.
- It ought to function in dangerous conditions, in the dark, and without issues with administration or motivation that would be profitable [48].
- The benefits should be maximized across a variety of application areas.
- It must be mobile, autonomous, and cognitive.

#### *2.3.2 Important qualities needed in construction robots*

Sensing and control, mobility and manipulation, human aspects and task factors, expert systems, and task flexibility are some of the qualities. The following sections describe these qualities in more detail.

##### *2.3.2.1 Sensing and control*

The largest challenge in developing construction-related robots is sensing and control, particularly in terms of navigation and position. The mobile autonomous robot requires location and heading data constantly for control. With the aid of video and image recognition systems, obstacles can be avoided and objects can be located. On numerous prototypes, obstacle avoidance is accomplished using touch sensors and ultrasonic technology [49].

##### *2.3.2.2 Mobility and manipulation*

The ability of equipment to move about construction sites is influenced by a number of variables, including the types of working environments and surface materials that must be traversed. For example, robots installed on rails have enough mobility to perform a variety of finishing activities and wall inspection jobs [50]. Different robots will do manipulative jobs according to their load bearing capabilities, arm length, and grip style.

*2.3.2.3 Human aspects and task factors*

The main motivation for the development of construction robots for use in severe settings, high and deep locations, boiling seas, and radiation zones is safety [51]. In these applications, the robots could operate alone or in tandem with a human operator working remotely from a secure location. Man or machine must have overall control for safety reasons. However, it is important to consider the current state of telecontrol to prevent delay issues when dealing with challenging tasks and to provide accurate manipulation feedback to the operator. During teleoperation, human factors are very crucial, even though automation reduces manual labor, it often requires more mental and cognitive effort. At this stage, the human-machine interface is particularly important for machine control and display.

*2.3.3 Expert system and task flexibility*

The construction industry will not benefit much from robotics on its own. Real progress can only be made when the construction process is completely organized. Expert systems, CAD/CAM, and database technologies are crucial in robots for task flexibility in civil engineering applications [52]. This field is now conducting in-depth research on unique programming environments, engineering graphics, logic, computation, and control requirements. The loadings, material properties, components, connectors, assembly, and geometric reasoning system all play a role in how construction components are represented in three dimensions by an expert system. A computerized work-control system will be necessary for robotic work to be employed efficiently. This system must include the construction site organization and sophisticated processing of commodities from the site to the robot.

## 2.4 Theoretical framework and hypotheses formulation

The Cognitive Behavioral Therapy (CBT), which was developed by psychologist Albert Ellis in the 1960s, and the theory of neural network, which was developed by Levin and Narendra in 1993, have been employed to guide this study. The principles of cognitive behavioral psychotherapies have described how a robo-mason can recognize, interact, think, and give meaning to situations, detect obstacles as well as form beliefs about themselves, their environments, and the world. This theory serves as a mechanism for robots to interact with its environments in an appropriate manner based on its cognitive behavior [53]. This theory better explains our study because the conventional robot design philosophy has given considerably more consideration to the embedded structure than to cognitive autonomous behavioral concerns, despite the fact that both play a substantial impact in the behaviors that follows. Autonomous cognitive design considerations of construction robo-mason are now essential in this study for a robot with embedded structure to learn and develop in order to eventually adapt to increasingly complicated building construction environments [54]. The major advantage is for the robot to change from dependent, excessive, and unhelpful behavior patterns to autonomous, advantageous, and balanced solutions.

The second theory is the theory of nonlinear dynamic system using neural network developed by Levin and Narendra [55]. The theory explains the use of neural networks to improve the stability and controllability of robotic systems. Though their study is restricted to nonlinear systems with complete state information access and feedforward MLNs with dynamic BP, their method takes into account a discrete-time

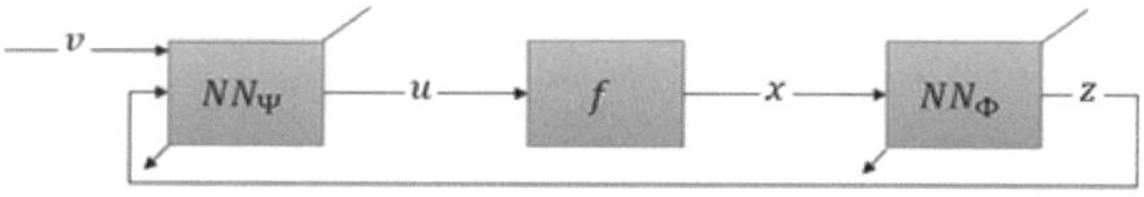


**Figure 9.**
*Proposed architecture of the neural networks in the work of Levin and Narendra [55].*

system at index k, as indicated in Eq. (1). The proposed design of the neural networks is shown in **Figure 9**.

$$\mathrm{x(k+1)=f(x(k),u(k))} \tag{1}$$

where $x(k) \in \chi \subset \mathrm{R}^n$, $u(k) \in U \subset \mathrm{R}^r$ and $f(0,0) = 0$ so that $x = 0$ is an equilibrium. Eq. (2) shows the conditions required for the neural networks to achieve feedback linearization and stabilization of the system.

$$\begin{cases} u = NN_\Psi(v,z) \\ z = NN_\varphi(x) \end{cases} (z,v) = \varepsilon R^n x R^r \tag{2}$$

However, Sontag tested and used this model to examine the potential and ultimate constraints of alternative neural networks designs [56]. He asserts that nonlinear systems like robots in general may be stabilized using neural networks with two hidden layers. Their conclusion seems to go against neural networks approximation theories, which contend that neural networks with a single hidden layer are the best approximators. In contrast to approximation problems, Sontag's solutions are based on the representation of the control problem as an inverse kinematics problem.

## 3. Method

### 3.1 Building deep learning algorithm for unsupervised robotics for construction industries

In building deep learning algorithm for unsupervised robotics for construction industries, we described as particular instances of a fairly simple recipe: combine a specification of a dataset, a cost function, an optimization procedure, and a model. For example, the linear regression algorithm combines a dataset consisting of $x$ and $y$, the cost function

$$\mathrm{J\,(w,b) = -E, p\ \log\ p\ (y|x)} \tag{3}$$

However, the model specification $p\ (y/x) = N\ (y;\ x\ w + b,\ 1)$, and, in most cases, the optimization algorithm defined by solving Eq. (3) for where the gradient of the cost is zero. By recognizing that these components can be replaced independently, a broad range of algorithms can be obtained. The cost function typically includes at least one term that causes the learning process to perform statistical estimation. Thus, the most common cost function is the negative log-likelihood, so that minimizing the cost function causes maximum likelihood estimation. The cost function may also include additional terms, such as regularization terms. For example, we can add weight decay to the linear regression cost function to obtain

$$J(w, b) = \lambda\|w\| - E,\ _{p} \log\ p\ (y|x). \tag{4}$$

This still allows closed-form optimization. For example, if we change the model to be nonlinear, then most cost functions can no longer be optimized in closed form. This requires us to choose an iterative numerical optimization procedure, such as gradient descent. The recipe for constructing a learning algorithm by combining models, costs, and optimization algorithms supports both supervised and unsupervised learning. The linear regression example shows how to support supervised learning. Unsupervised learning can be supported by defining a dataset that contains only $x$ and providing an appropriate unsupervised cost and model. However, we can obtain the first PCA vector by specifying that our loss function is

$$J(w) = E,\ _{p}\ \|x - r(x; w)\| \tag{5}$$

while our model is defined to have $w$ with norm one and reconstruction function $r(x) = w\ xw$. In some cases, the cost function may be a function that we cannot actually evaluate, for computational reasons. In these cases, we can still approximately minimize it using iterative numerical optimization so long as we have some way of approximating its gradients. Most machine learning algorithms make use of this recipe, though it may not immediately be obvious. If a machine learning algorithm seems especially unique or hand-designed, it can usually be understood as using a special-case optimizer. Some models such as decision trees or k-means require special-case optimizers because their cost functions have flat regions that make them inappropriate for minimization by gradient-based optimizers. Recognizing that most machine learning algorithms can be described using this recipe helps to see the different algorithms as part of a taxonomy of methods for doing related tasks that work for similar reasons, rather than as a long list of algorithms that have separate justifications.

## 4. Discussion

These feature learning approaches are one of the major strengths of modern deep learning methods. Since these algorithms are able to learn good features from data, they are much less sensitive to input representations than other conventional learning algorithms such as support vector machines, Gaussian processes, and others. Deep learning algorithms are able to learn good representations and solve problems even from basic representations such as raw pixels, avoiding the need to hand-design features as with other learning algorithms, saving significant engineering effort for many of the complex problems encountered in robotics, where features can be unintuitive and hard to design.

## 5. Conclusion

The branch of computer science that is now permeating the construction sector is robotics. It is crucial to automate construction sites so that robots can perform risky tasks for workers in dangerous environments like high altitudes, deep water, high radiation zones, inclement weather, and deep oceans. It is also beneficial in terms of avoiding the disruptive effects of strikes, issues with administration and motivation, safety and health regulations, a lack of skilled labor, and the need to perform repetitive, dirty, and dangerous work as well as the completion of projects or tasks with quality control, on schedule, and economically. Despite the fact that this technology is helping the construction industry, much research is still needed in the areas of sensing and control, human factors, task flexibility, and the software support to integrate robots into a larger construction-based management. Our feature learning approach is one of the major strengths to achieve this. Since this algorithm is able to learn good features from data, they are much less sensitive to input representations than other conventional learning algorithms such as support vector machines, Gaussian processes, and others. Our deep learning algorithm is able to learn good representations and solve problems even from basic representations such as raw pixels, avoiding the need to hand-design features as with other learning algorithms, saving significant engineering effort for many of the complex problems encountered in robotics, where features can be unintuitive and hard to design.

## Acknowledgements

The authors express their gratitude to all that helped in the actualization of this research.

## Conflict of interest

No potential conflict of interest was reported by the author(s).

## Author details

Nicholas Eze[1*], Ekene Ozioko[2] and Johnpaul Nwigwe[2]

1 University of Nigeria, Nsukka, Nigeria

2 Enugu State University of Science and Technology, Enugu, Nigeria

*Address all correspondence to: nicholas.eze@unn.edu.ng

## References

[1] Castro-Lacouture D. Construction automation. In: Nof SY, editor. Springer Handbook of Automation. Berlin: Springer; 2009. pp. 1063-1078. DOI: 10.1007/978-3-540-78831-7

[2] Vähä P, Heikkilä T, Kilpeläinen P, Järviluoma M, Gambao E. Extending automation of building construction—Survey on potential sensor technologies and robotic applications. Automation in Construction. 2013;**36**:168-178

[3] Mahajan G. Applications of drone Technology in Construction Industry: A study 2012-2021. International Journal of Engineering and Advanced Technology. 2021;**11**(1):224-239. DOI: 10.35940/ijeat.A3165.1011121

[4] De Blois M, Lizarralde G, De Coninck P. Iterative project processes within temporary multi-organizations in construction: The self-, eco-ReOrganizing projects. Project Management Journal. 2016;**47**(1):27-44. DOI: 10.1002/pmj.21560

[5] Navon A, Kelly K, Johnston J. Human factors in introducing on-site construction automation. Journal of Construction Engineering and Management. 1993;**119**(4). DOI: https://doi.org/10.1061/(ASCE)0733-9364(1993)119:4(801)

[6] Elghaish F, Matarneh S, Talebi S, Kagioglou M, Hosseini MR, Abrishami S. Toward digitalization in the construction industry with immersive and drones technologies: A critical literature review. Smart and Sustainable Built Environment. 2021;**10**(3):345-363. DOI: 10.1108/SASBE-06-2020-0077

[7] Hussain CM, Samiha N, Paulraj MS. Source reduction and waste minimization. In: Advanced Zero Waste Tools: Present and Emerging Waste Management Practices. Vol. 2. London: Elsevier; 2021. pp. 1-22. DOI: 10.1016/C2020-0-01110-2

[8] Shukla A, Karki H. Application of robotics in onshore oil and gas industry—A review part I. Robotics and Autonomous Systems. 2015;**75**:490-507

[9] Oh J-K, Jang G, Oh S, Lee JHJS, Yi B-J, Moon YS, et al. Bridge inspection robot system with machine vision. Automation in Construction. 2009;**18**:929-941. DOI: 10.1016/j.autcon.2009.04.003

[10] Albeaino G, Gheisari M, Franz BW. A systematic review of unmanned aerial vehicle application areas and technologies in the AEC domain. Journal of Information Technology in Construction. 2019;**24**:381-405

[11] Richard R-B. Industrialised building systems: Reproduction before automation and robotics. Automation in Construction. 2005;**14**:442-451. DOI: 10.1016/j.autcon.2004.09.009

[12] Bock T. The future of construction automation: Technological disruption and the upcoming ubiquity of robotics. Automation in Construction. 2015;**59**:113-121. DOI: 10.1016/j.autcon.2015.07.022

[13] Saeed BN. Introduction to Robotics, Analysis, Control, Application. India: Willey Publications; 2017

[14] Schmidhuber J. Curious model-building control systems. In: Proceedings of the International Joint Conference on Neural Networks. Vol. 2. Singapore: IEEE; 1991. pp. 1458-1463

[15] Herrmann J, Pawelzik K, Geisel T. Learning predictive representations. Neurocomputing. 2000;**32-33**:785-791

[16] Thrun S. Exploration in active learning. In: Arbib M, editor. Handbook of Brain Science and Neural Networks. Massachusetts, London: Cambridge Press; 1995

[17] Rosenblatt F. Principles of Neurodynamics. New York: Spartan; 1962

[18] Minsky M, Papert S. Perceptrons. Cambridge; Massachusetts: MIT Press; 1969

[19] Rumelhart DE, McClelland JL, editors. Parallel Distributed Processing. Cambridge; Massachusetts: MIT Press; 1986

[20] Zhang S, Yao L, Sun A, Tay Y. Deep learning based recommender system: A survey and new perspectives. ACM Computing Surveys (CSUR). 2019; **52**(1):5

[21] Najafabadi MM, Villanustre F, Khoshgoftaar TM, Seliya N, Wald R, Muharemagic E. Deep learning applications and challenges in big data analytics. Journal of Big Data. 2015;**2**:1

[22] Prieto A, Prieto B, Ortigosa EM, Ros E, Pelayo F, Ortega J, et al. Neural networks: An overview of early research, current frameworks and new challenges. Neurocomputing. 2016;**214**(19):242-268

[23] Wehle H-D. Machine learning, deep learning, and AI: What's the difference? In: Conference Paper July 2017. Germany: New Generation Books. 2017

[24] Lorenc SJ, Handlon BE, Bernold LE. Development of a robotic bridge maintenance system. Automation in Construction. 2000;**9**:251-258. DOI: 10.1016/S0926-5805(99)00040-0

[25] Ajayi OG, Salubi AA, Angbas AF, Odigure MG. Generation of accurate digital elevation models from UAV acquired low percentage overlapping images. International Journal of Remote Sensors. 2017;**38**(8–10):3113-3134

[26] Allasia P, Baldo M, Giordan D, Godone D, Wrzesniaj A, Lollino G. Near real time monitoring systems and periodic surveys using a multisensory UAV: The case of Ponzano landslide. In: IAEG/AEG Annual Meeting Proceedings. Vol. 1. San Francisco, California: Cham: Springer; 2018. pp. 303-310

[27] Mary N, Shafiyia S, Ben Maaouia M. Applications of drone Technology in Construction Projects: A systematic literature review. International Journal of Research - Granthaalayah. 2022; **10**(10):1-14. DOI: 10.29121/granthaalayah.v10.i10.2022.4810

[28] Jenson J. An introduction to UAV Photogrammetry software. 2019. Available from: https://uavcoach.com

[29] Aicardi I, Nyapwere N, Nex F, Gerke M, Lingua AM, Koeva MN. Coregistration of multi emporal UAV image dataset for monitoring applications: A new approach. International Archive of the photogrammetry, Remote Sensing and Spatial Information Sciences. 2016;**XLI-B1**:757-763

[30] Siegwart R, Nourbakhsh IR, Scaramussa D. Introduction to Autonomous Mobile Robots. Cambridge, Massachusetts London, England: MIT Press; 2004

[31] Giordan D, Adams MS, Aicardi I, et al. The use of unmanned Areial vehicles (UAV) for engineering geology applications. Bulletin of Engineering Geology and the Environment. 2020;**79**: 3437-3481. DOI: 10.1007/s10064-020-01766-2

[32] De Melo RRS, Costa DB, Alvares JS, Irizzary J. Applicability of unmanned aerial system (UAS) for safety inspection on construction site. Safety Science. 2017;**98**:174-185. DOI: 10.1016/j.ssci 2017.06.008

[33] Vieri GS, Gianluca B, Emilio C. Autonomous Reinforcement Learning of Multiple Interrelated Tasks. IEEE; 2019

[34] Rajkumar R, Roshini C, Sadhana S, Saranya S. Robotics in civil engineering. International Journal of Scientific & Engineering Research. 2017;**8**:10

[35] Screpanti L, Miotti B, Monteriu A. Robotics in education: A smart and innovative approach to the Chalenges of the 21$^{st}$ century. In: Scaradozzi D, Guasti L, Di Stasio M, Miotti B, Monteriu A, Blikstein P. (editors). Makers at School, Educational Robotics and Innovative Learning Environments. Lecture Notes in Networks and Systems. 2021; 240. Springer, Cham. doi: 10.1007/ 978-3-030-77040-2_3

[36] Attuquayefio S, Addo H. Review of obstacles which inhibit ICT adoption in higher education. European Scientific Journal. 2014;**10**(13):1857-7881

[37] Chauhi BA, Chausi A, Dka Z. ICT for higher education: An outlook on the cost saving IT projects and information systems. Conference: Managing intellectual capital and innovation for sustainable and inclusive society. In: Proceedings of the MakeLearn and TIIM Joint International Conference. Bari, Italy: International Academic Publisher; 2015. p. 63

[38] Odia LO, Omofonmwan SI. Educational system in Nigeria problems and prospects. Journal of Social Sciences. 2007;**14**(1):86-85. DOI: 10.1080/ 09718923.2007.11978347

[39] Gogu G. Structural Synthesis of Parallel Robots—Part 5: Basic over-Constrained Topologies with schönflies Motions. New York: Springer; 2014

[40] Hess J, Weinstein M, Welch L. Ergonomic best practices in masonry: Regional differences, benefits, barriers, and recommendations for dissemination. Journal of Occupational and Environmental Hygiene. 2010;7:446-455. DOI: 10.1080/15459624.2010.484795

[41] Vink P, Miedema M, Koningsveld E, van der Molen H. Physical effects of new devices for bricklayers. International Journal of Occupational Safety and Ergonomics, Taylor & Francis. 2002;**8**: 71-82. DOI: 10.1080/ 10803548.2002.11076515

[42] Pritschow G, Dalacker M, Kurz J, Zeiher J. A mobile robot for on-site construction of masonry. In: Proceedings of IEEE/RSJ International Conference on Intelligent Robots and Systems (IROS'94). Vol. 3. Piscataway, NJ: IEEE; 1994, 1701–1707. DOI: 10.1109/IROS.1994.407628

[43] Tan N, Mohan RE, Watanabe A. Toward a framework for robot-inclusive environments. Automation in Construction. 2016;**69**:68-78. DOI: 10.1016/j.autcon.2016.06.001

[44] Pivac M, Pivac M. Fastbrick robotics. 2016. Available from: https:// www.fbr.com.au/

[45] Podkaminer N, Peters L. S. Construction robotics. 2015. Available from: https://construction-robotics.com/

[46] Pritschow G, Dalacker M, Kurz J, Gaenssle M. Technological aspects in the development of a mobile bricklaying robot. Automation in Construction. 1996;5:3-13. DOI: 10.1016/0926-5805 (95)00015-1

[47] Seward N, Bonev I. A new 6-dof parallel robot with simple kinematic model. In: Proceedings of the 2014 IEEE International Conference on Robotics and Automation (ICRA 2014). Hong Kong; China: IEEE; 2014

[48] Eze NU, Obichukwu PU, Kesharwani S. Perceived usefulness, perceived ease of use in ICT support and use for teachers. IETE Journal of Education. 2021;**62**:1. DOI: 10.1080/09747338.2021.1908177

[49] Abayomi OA, Olaitan A, Odiete JO, Olujide AA. An autonomous obstacle avoidance robot using ultrasonic sensor. Journal of Computer Science and Its Application. 2020;**27**(1):68-71. DOI: 10.4314/jcsia.v27i1.15

[50] Brock O, Park J, Toussaint M. Mobility and manipulation. In: Siciliano B, Khatib O, editors. Springer Handbook of Robots. Springer, Cham: Springer Handbooks; 2016. DOI: 10.1007/978-3-319-32552-1_40

[51] Onososen AO, Musonda I, Ramabodu M. Construction robotics and human–robot teams research methods. Buildings. 2022;**12**:1192. DOI: 10.3390/buildings12081192

[52] Borja G, Miroslaw JS. Future of robotics and automation in construction. Chapter 15. In: Future of Robotics and Automation in Construction. London: Routledge; 2020. DOI: 10.1201/9780429398100-15

[53] Daily M, Cho Y, Martin K, Payton DW. World embedded interfaces for human-robot interaction. In: Proceedings of the 36th Annual Hawaii International Conference on System Sciences. Hawaii, USA: IEEE; 2003. DOI: 10.1109/HICSS.2003.1174285

[54] Dubor A, Izard J, Cabay E, Rodriguez M, Sollazzo A, Markkopoulou A. On-site robotics for sustainable construction: Foreword by Sigrid Brell-Çokcan and Johannes Braumann, Association for Robots in architecture. In: Robotic Fabrication in Architecture, Art and Design. Cham: Springer Science & Business Media; 2018. DOI: 10.1007/978-3-319-92294-2_30

[55] Levin AU, Narendra KS. Control of nonlinear dynamical systems using neural networks: Controllability and stabilization. IEEE Transactions on Neural Network. 1993;**4**:192-206

[56] Sontag ED. Feedback stabilization using two-hidden-layer nets. IEEE Transactions on Neural Network. 1992;**3**:981-990

Chapter 4

# The Use of Artificial Intelligence to Bridge Multiple Narratives in Disaster Response

*Karla Saldaña Ochoa*


## Abstract

Disaster response presents the current situation, creates a summary of available information on the disaster, and sets the path for recovery and reconstruction. During the last 10 years, various disciplines have investigated disaster response twofold. First, researchers published several studies using state-of-the-art technologies for disaster response. Second, humanitarian organizations have produced numerous mission statements on how to respond to natural disasters. The former suggests questioning: If we have developed a considerable number of studies to respond to a natural disaster, how can we cross-validate its results with humanitarian organizations' mission statements to bring the knowledge of specific disciplines into disaster response? To address the above question, the research proposes an experiment that considers both: knowledge produced in the form of 8364 abstracts of academic writing on Disaster Response and 1930 humanitarian organizations' mission statements indexed online. The experiment uses an Artificial Intelligence algorithm, Neural Network, to perform the task of word embedding—Word2Vec—and an unsupervised machine learning algorithm for clustering—Self-Organizing Maps. Finally, it employs Human Intelligence for the selection of information and decision-making. The result is a digital infrastructure that will suggest actions and tools relevant to a specific scenario, providing valuable information loaded with architectural knowledge to guide the decision-makers at the operational level in tasks dealing with spatial and constructive constraints.


**Keywords:** artificial intelligence, disaster response, big data, Word2Vec, self-organizing maps, architecture

## 1. Introduction

After a natural disaster, disaster response is essential. Its assessments present the current situation and summarize information that guides rescue forces and other immediate relief efforts to the site [1]. Currently, many scholars publish research on disaster response, focusing on implementing artificial intelligence to process big data. Examples of the former are: a platform to classify crisis-related communications [2], a model that learns damage assessment and proposes applications [1], an automated method to retrieve information for humanitarian crises [3], or an interface to assist with fast decision-making during humanitarian health crises [4, 5]. In parallel, we observe a growing global presence of humanitarian organizations that publish humanitarian mission statements and declarations. There are more than 4000 organizations active in this area. Examples of large ones include the World Food Program (WFP), Cooperative Assistance and Relief Everywhere (CARE), the International Federation of Red Cross and Red Crescent Societies (IFRC), and Action Against Hunger (AAH) [6]. However, the two narratives do not overlap and often ignore each other's efforts. To achieve an understanding that leads to a unification of the knowledge produced, one must think of methods to analyze both academic literature and humanitarian mission statements to create a common ground.

Most scientific publications and mission statements of humanitarian organizations are published in text form, which is challenging to analyze using traditional statistical analysis. Nowadays, artificial intelligence algorithms are used to facilitate this process. The branch of research that concentrates on text analysis with these methods is called text mining—a specific branch of data mining. Text mining is the process of analyzing texts to obtain new information from them. It does this by identifying patterns or correlations between terms, thus making it possible to find information that is not explicit in the text. This new information makes it possible to create inferences in different tasks such as classification, clustering, or prediction of texts. Currently, novel studies strengthen their efforts to develop or improve methods for analyzing text data, such as the study by [7], which will serve as a reference for this experiment.

## 2. Methodology

This article will focus on the late operations in disaster response: building safety assessments, temporary housing, and policy recommendations. The abovementioned tasks will serve as keywords to crawl academic publications and humanitarian organizations' mission statements to define a specific framework directed toward architectural practices to respond to natural disasters.

### 2.1 Academic publications

To collect academic publications, the pipeline follows the approach of [8]. In their article, the researchers proposed to work with abstracts. They argue that abstracts communicate information concisely and straightforwardly, avoiding unnecessary words. In contrast with full texts, which sometimes contain negative relationships, writing styles can include complex sentences that require different encoding methods than the ones proposed in the present research.

The abstracts of academic publications will be collected from the ScienceDirect database (https://www.sciencedirect.com), one of the most significant and extensive academic archives of journals and conference proceedings. Using a diverse database such as ScienceDirect is beneficial, as a substantial amount of scientific literature from many fields can be assembled to address architectural practices in disaster response. To start collecting the abstracts of academic publications, an API was used, which required keywords to filter data according to a specific interest. The API has two syntax rules. The first rule, +AND+, searches for pair words, the second, +, determines that the word must be contained in the text. By setting the keywords: Bui lding+AND + Safety+Assessments, Housing+and + Disaster+ANDResponse, and Poli cy+AND + Recommendations+and + Disaster+AND + Response, 9000 abstracts were collected. The articles are from the last 10 years, from 2010 to 2020.

As suggested by Tsitoyana et al. [8], abstracts that contained in their titles the following keywords: "Foreword," "Prelude," "Commentary," "Workshop," "Conference," "Symposium," "Comment," "Retract," "Correction," "Erratum," and "Memorial" were removed. **Figure 1** shows a word cloud of the collected abstracts, where the main concepts are represented, and an overview of the general intention is graspable.

## 2.2 Professional humanitarian organizations

Professional Humanitarian Organizations have continuously expanded over the past decades. Worldwide more than 125 million people relying on humanitarian aid—double the number 10 years ago [7]. Commonly, such organizations communicate with their stakeholders via platforms online, sharing their mission statements, aims, and goals. Two of the leading online platforms are Wikipedia and ReliefWeb. Hence, data on humanitarian organizations' mission statements will be collected from these two web sources.

From Wikipedia, 749 humanitarian mission statements were collected using Wikipedia API under the Category "Humanitarian aid organizations." The collected texts were translated into English. To collect the text from ReliefWeb, a twofold

**Figure 1.**
*World cloud of 8364 abstracts from the field of disaster response.*

process was performed. The first step was to collect 1154 links of Humanitarian Organizations from the ReliefWeb web (https://reliefweb.int), a specialized digital service of the UN Office for the Coordination of Humanitarian Affairs (OCHA). Those links were organized under the tag "Organizations." The second step was to access each retrieved link and parse the source text, searching for tags: "about," "about us," "we are," "who we are," and "what we do." Having a particular focus on English texts. After the two-step process, a description of each organization and mission statement was found.

After joining both datasets (Wikipedia and ReliefWeb), an overlapping of 87 organizations was found. This points out the lacking of consistency regarding professional humanitarian aid on the web. Nevertheless, we worked with all the humanitarian organizations found, which sum up 1930 entries. **Figure 2** shows the word cloud of the overall dataset.

### 2.3 Data processing and encoding

To pre-process the text data, three operations were implemented: first, selection of lower-casing and de-accenting; second, removing stop words; and third, selection of words that are part of a speech (nouns, pronouns, adjectives, verbs, adverbs, prepositions, conjunctions, and interjections).

**Figure 2.**
*World cloud of 1930 mission statements from humanitarian organizations.*

To encode the text data to numerical vectors, the research uses algorithms for word embedding (Natural Language Processing), a technique that assigns high-dimensional vectors (embeddings) to words in a text corpus, preserving their syntax and semantics. Tshitoyan et al. [8] demonstrated that scientific knowledge could efficiently be encoded as information-dense word embeddings without human labeling or supervision. The algorithm used to transform the text into word embeddings is an Artificial Neural Network called Word2Vec [9, 10], which uses the continuous-bag-of-words (CBOW) method. The algorithm learns the embedding by maximizing the ability of each word to be predicted from its set of context words using vector similarity. The output of Word2Vec is a 50-dimensional numerical vector for each word from the text corpus.

To continue with the experiment, a subsample of 40% of the pre-processed texts (both abstracts of academic writing and mission statements of humanitarian organizations) were fed as training data to a Word2Vec algorithm to use the knowledge acquired from the previous training to create a domain-specific model Word2VecDR (this is call transfer learning). After the Word2VecDR was trained, it was able to encode all texts from the dataset into a numerical representation—every word of the text was assigned 50 numerical vectors. The texts ranged from 15 to 5668 words, having an average of 824 words per text. Therefore, if an abstract contains 100 words, the resulting vector form Word2VecDR is a list of 100 sub-lists with 50 elements each.

Mikolov et al. [9] observed that simple algebraic operations on word embeddings, e.g., vector "King" – vector "Man" + vector "Woman," results in a vector that is closest to the vector "Queen" concluding that the resulting vector is content-related. Furthermore, researchers have also applied statistical operations such as mean or average values onto a list of word embeddings and had successful results that captured the content of the text (examples of the former can be found in [11, 12]. However, when calculating the mean value or adding each word vector, the resulting vector will be an abstraction (reduced) of its content, hence, losing information. To encapsulate as much information as possible from the list of numerical vectors, the present research proposed to use Higher-Order Statics (HOS). In HOS, mean (X) and standard deviations (s) are related to the first and the second-order moments—one could calculate up to n-order moments. Skewness (ski) can be calculated from the third-order moments of the data distribution, which measures the direction of the tail in comparison to the normal distribution, where Y is the Median.

$$sk_i = \frac{(X - Y)}{s} \tag{1}$$

If the resulting number is positive, the data are skewed to the left, leaving the tail pointing to the right side of the distribution. If the resulting number is negative, the tail is on the left side of the distribution. Kurtosis (ki) is the fourth-order moment, which measures how heavy the tails of a distribution are [13], where N is the sample size.

$$k_i = \frac{\sum_{i=1}^{N} \frac{(X_i - X)}{N}}{s^4} \tag{2}$$

By applying the fourth moments of HOS to the data, each text is represented by a numerical vector of 200 dimensions or four sublists of 50 dimensions for each HOS moment (mean, standard deviation, skewness, and kurtosis). Two advantages of encoding data with HOS are that, compared with the embedding vectors in deep autoencoders, the resulting vectors of HOS are meaningful and directly interpretable [14]. Second, by using HOS, the computational time for clustering the text data reduces exponentially since the length of the numerical vector is decreased. Additionally, by using the fourth moments of HOS, each resulting numerical vector encapsulates more information than when using only one statistical value (first or second moment).

### 2.4 Data representation and clustering

Clustering and representation techniques assist with adequately exploring data collection and identifying clusters of similar information that share similar properties [15]. For example, to cluster text, the following algorithms have been used: Support Vector Machine (SVM) [16], k-means [17], Principal Component Analysis (PCA) [18], and Kohonen Self-Organizing Map (SOM) [19]. A full review of different clustering algorithms can be found in [20]. As shown in the work of [19], the unsupervised ML algorithm SOM [21, 22] has proven to have excellent performance when clustering text data and reducing its dimensionality. Additionally, as presented in the work of [23].

> *"SOM acts as a nonlinear data transformation in which data from a high-dimensional space are transformed into a low-dimensional space (usually a space of two or three dimensions), while the topology of the original high-dimensional space is preserved. SOM has the advantage of delivering two-dimensional maps that visualizes data clusters that reflect the topology of the original high-dimensional space."*

In the proposed experiment, the algorithm of choice for Clustering is SOM, which takes advantage of both clustering and dimensionality reduction [21, 22]. SOM acts as a nonlinear data transformation in which data from a high-dimensional space are transformed into a low-dimensional space (usually a space of two or three dimensions) while preserving the topology of the original high-dimensional space. Topology preservation means that if two data points are similar in the high-dimensional space, they are necessarily close in the new low-dimensional space, and hence, they are placed within the same cluster. This low-dimensional space, usually represented by a planar grid with a fixed number of points, is called a map. Each node of this map has specific coordinates (xi,1, xi,2) and an associated n-dimensional vector or Best Matching Unit (BMU) in such a way that similar data points in the high dimensional space are given similar coordinates. Moreover, each node of the map represents the average of the n-dimensional original observed data that, after iteration, belongs to this node [14]. SOM is an algorithm that helps navigates a dataset. It represents the many dimensions of a dataset into one or two dimensions, allowing a deep understanding of such a dataset.

## 3. Results

At this point, the text of the academic publications and the humanitarian mission statements are comparable since both were encoded with the same method (Word2Vec). Therefore, they can be clustered using the SOM algorithm presented in the previous section. The first attempts to unify the narratives showed that the number of texts still creates confusion, as many studies and mission statements do not focus specifically on architectural practices in disaster response. Instead, such clustering has a general approach, making it challenging to select studies or tools to answer questions about spatial or constructive problems. Therefore, we will use the SOM as a first filter to narrow the focus to specific tasks (architectural practices). The SOM algorithm allows us to have an organized view of the data; that is to say, we can make a selection based on a specific approach. Thus, it will enable us to concentrate on the information relevant to the focus of this experiment.

The filtering will be performed on both datasets beginning with the mission statements of humanitarian organizations. For this purpose, the numerical vectors representing the texts from humanitarian organizations were fed as inputs in a 10x10 SOM grid. The algorithm started with an initial distribution of random weights, and over

**Figure 3.**
*The SOM HO is a SOM grid of 10×10 data from the humanitarian organization data.*

1 million epochs eventually settled into a map of stable zones or clusters. The output layer can be visualized as a smooth changing spectrum where each SOM node has its coordinates and an associated n-dimensional vector or Best Matching Unit (BMU). For visualization purposes, a color assigned to the weight value (n-dimensional vector) of each BMU and a list of keywords are displayed together. The keywords are the most common terms used in the texts that are clustered in each SOM node. The size of the word represents the number of occasions the word appeared in the group of text. **Figure 3** shows the consistency in clustering, such as similar keywords being positioned close to each other. This trained SOM grid can be considered a common ground for the mission statements of humanitarian organizations, which will be called SOM HO.

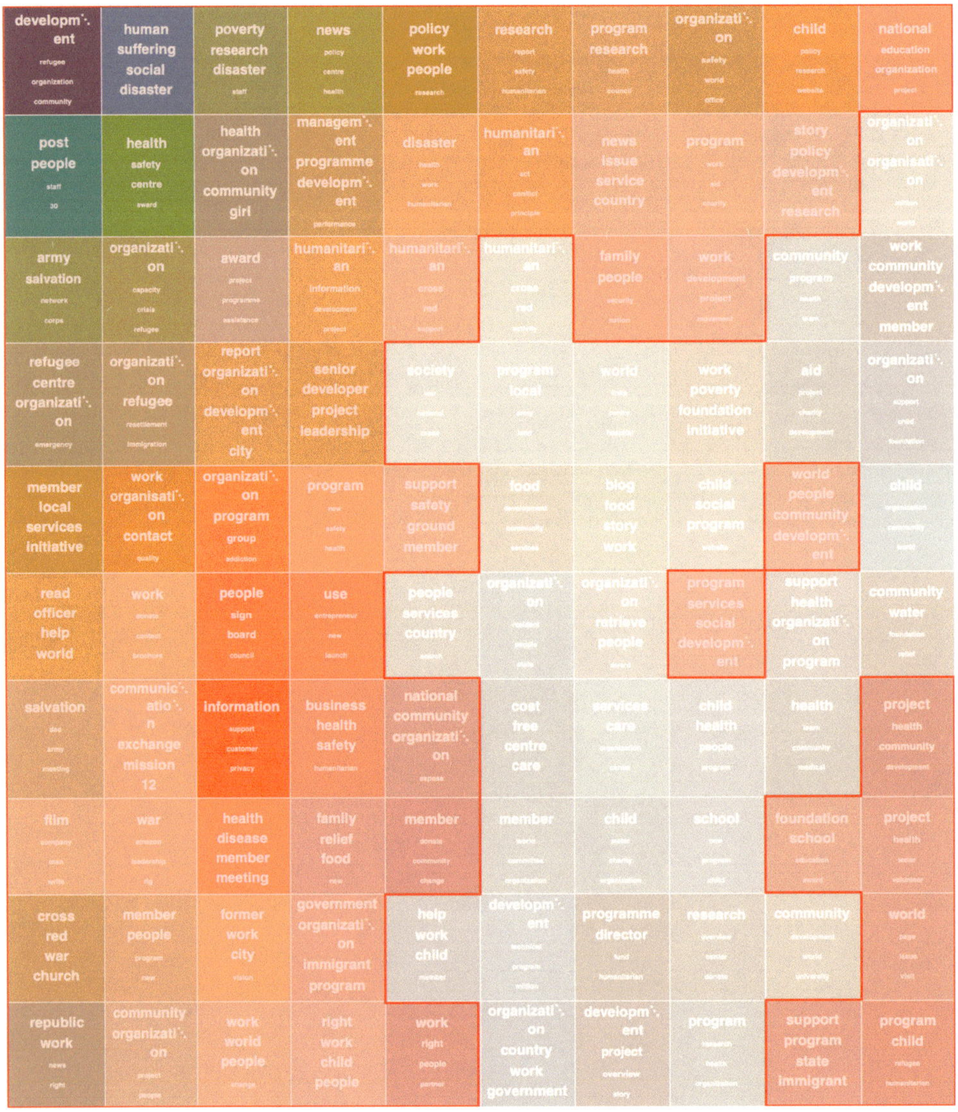


**Figure 4.**
*Shows the activated cells of the SOM HO, where humanitarian organizations that share a common interest in academic writing are found.*

Although this SOM HO represents data in an organized manner, it also contains data on humanitarian organizations that may not be relevant to this thesis's focus. As our focus is on architectural practices for natural disaster response, we will perform filtering of the humanitarian organization's data based on similar interests shared with academic publications. When a new dataset is fed to a trained SOM, each unique data point measures its Euclidean distance to each BMU of the trained SOM. The closer the distance, the better the data point fits that node. When feeding the data on academic literature into the trained SOM HO, this data activates a specific number of cells. Hence, the data we will take for creating our final common ground comes only from the cells activated by the academic publication data. **Figure 4** shows the activated cells of the SOM HO, where humanitarian organizations that share a common interest in academic writing are found.

After collecting all the activated cells, 1081 humanitarian organizations out of the original list of 1930 were filtered. Some of the selected humanitarian organizations are: The International NGO Safety Organization, Nansen International Office for Refugees, Peoplesafe, Rise Against Hunger, SeedChange, and Association of Assistance Solidarity Supportiveness of Refugees and Asylum Seekers.

A similar filtering process was performed on the data on academic writing—initially 8364. First, the encoded abstracts were fed into a new 2D SOM algorithm of 10×10, where each text was clustered based on the similarity of content (the training procedure was like that in the case of humanitarian organization mission statements)—creating a SOM of academic publications, which we will call SOM AP. The filtering constraint was defined by a sample of publications that will be taken as exemplary text. Those samples of academic publications concentrate on architectural practices for disaster response. Such selected academic publications were fed into the trained SOM AP and found some cells with similar approaches. Out of the 8364 abstracts, 835 were selected. **Figure 5** shows the Map of Events AP and the selected cells matching the selected publications.

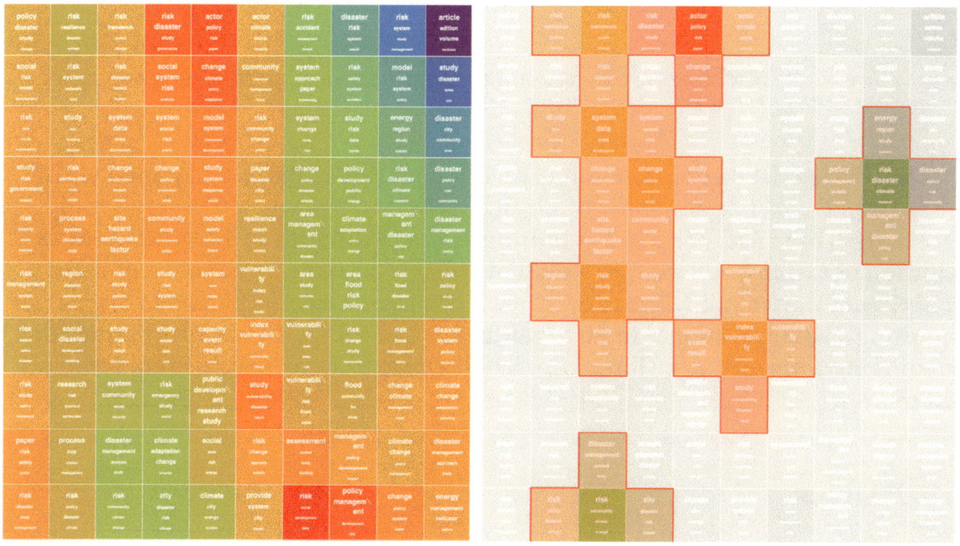

**Figure 5.**
*SOM 10×10 of abstracts and the selected cells based on specific literature. Resulting in the final filtering of 835 abstracts.*

**Figure 6.**
*Common Ground, a SOM of 10x10 trained with joined data from filtered humanitarian organizations (1082) and academic writing from the field of disaster response (835).*

The filtered dataset of humanitarian organizations (1082) was joined with the filtered dataset of academic publications (835), creating a new dataset of 1917 texts. These data were fed as input in a new SOM grid of 10×10 that, after a million iterations, settled into a map of clustered texts or what we will call a final Common Ground (**Figure 6**). Such a Common Ground joins two discourses regarding disaster response in an organized manner that serves as a ground for articulating informed decisions, which will emerge out of specific requirements and interests.

## 4. Validation

To validate the proposed approach, we will take international news (reports on natural disasters and their impact on the communities) since, after a natural disaster, the first source of information comes from that source. It should be emphasized that any query can be taken, being this something very particular or something descriptive, as in the case of the news. For experimental purposes, the specific interest (query) was extracted from news describing three natural disasters of the last 5 years.

The first one was the 2016 Earthquake of magnitude 7.8 on the Ecuadorian coast:

> *"A magnitude 7.8 earthquake rocked Ecuador's coast on April 16, 2016 — killing almost 700 people and leveling homes, schools, and infrastructure. More than 6,000 people were severely injured. The quake's epicenter was offshore, about 17 miles from the town of Muisne in Manabí province and 100 miles northwest of Quito, the capital. After the quake, more than 700,000 people needed assistance. An estimated 35,000 houses were destroyed or badly damaged, leaving more than 100,000 people in need of shelter. Water, sanitation, and healthcare facilities were also destroyed."* [21, 22]

The second one from the 2019 earthquake of 5.4 magnitude in Costa Rica.

> *"A magnitude 5.4 earthquake shook much of Costa Rica at 7:33 p.m., according to preliminary data from the National Seismological Network (RSN). The tremor had an epicenter near Arancibia, Puntarenas, which is located about 45 miles northwest of San José and its surrounding Greater Metropolitan Area. RSN indicates the quake was felt throughout the Central Valley, home to nearly three-quarters of Costa Rica's population. There have not been any immediate reports of substantial damage. "According to preliminary data from the emergency committees, so far there is no report of damage after the perceived earthquake, "said the National Emergency Commission (CNE) in a post. The National Seismological Network has already reported at least one aftershock, which occurred at 7:38 p.m. and had a similar epicenter."* [23]

The third one from the 2019 earthquake with a magnitude of 6.5 in Indonesia.

> *"A 6.5-magnitude earthquake struck the remote Maluku Islands in eastern Indonesia on Thursday morning, killing at least 20 people. Indonesian officials said the quake, which was detected at 8:46 a.m. local time, did not present the threat of a tsunami. But it was classified as a "strong" earthquake in Ambon, a city of more than 300,000 people and the capital of Maluku Province. The United States Geological Survey said the epicenter was about 23 miles' northeast of Ambon. At least 20 people were killed in the quake, the authorities said, including a man who was killed when a building partially collapsed at an Islamic university in Ambon, according to Reuters. More than 100 people were reported injured in the quake, and the authorities said about 2,000 had been displaced from their homes."* [24]

All queries were pre-processed and fed into the trained Word2Vec model to extract its word embeddings. The output of the Word2Vec model was encoded with HOS (see section Data Processing), having a final vector of 200 dimensions for each query. Each query vector was then fed into the final Common Ground, SOM of 10×10, (**Figure** 7), finding the closest Euclidean distance to a BMU. **Figure** 7 displays the closest BMU concerning each query vector. For the first case study, the keywords in the selected cell are: community, earthquake, risk, and safety. For the second case study, the keywords in the selected cell are: disaster, community, study, and management. For the third case study, the keywords in the selected cell are: energy, disaster, policy, study.

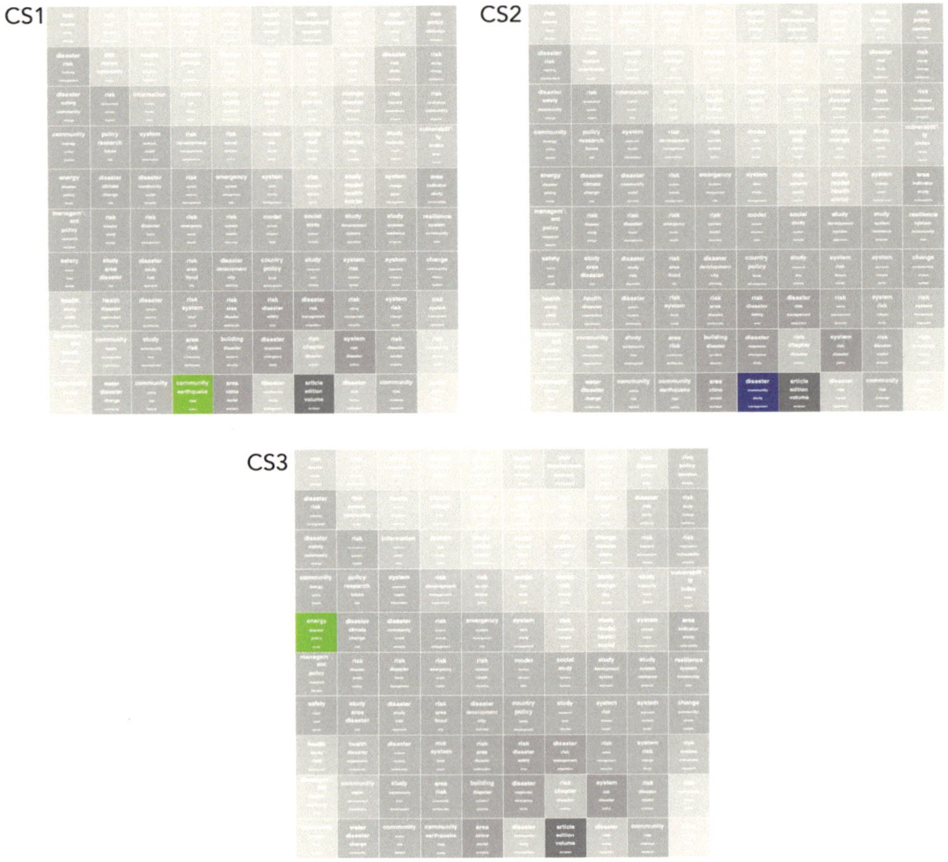


**Figure 7.**
*On color, the BMU had the closest Euclidean distance to each query vector.*

## 5. Discussion

As explained in the Introduction, this research focuses on operations for disaster response, where information has to be concise and arrives on time for decision-makers at the operational level to make an informed decision. The selected case studies are only a subsample of the different scenarios investigated in this work. The final output for each query was four organizations, and four studies were selected (**Figure 8**). These texts belong to the set of texts grouped in the BMU assigned to each query.

A correlation with the specific interest can be observed when analyzing the final output. Let us focus on the first case study. Here we can see that the keywords extracted from the first query (2016 Earthquake of magnitude 7.8 on the Ecuadorian coast) are: earthquake, people, severely, assistance, shelter, health care, facility; and that those assigned to the BMU were: community, earthquake, risk, safety; have several overlapping, which demonstrates the consistency in the clustering. Within the selected cell were various humanitarian organizations and academic publications from different research fields such as health sciences, engineering, and architecture. These are all clustered together as they share similar keywords and concerns. This pattern will always occur; an output will always suggest a list of possible options from different fields of study. Therefore, the final selection, which can be called human

CS1 **Tools**

The role of post-disaster public policy responses in housing recovery of tenants
www.sciencedirect.com/science/article/pii/S0197397513000489

Anticipated behavioral response patterns to an earthquake: The role of personal and household characteristics, risk perception, previous experience and preparedness
www.sciencedirect.com/science/article/pii/S2212420918300244

Seismic response of shield tunnel subjected to spatially varying earthquake ground motions
https://www.sciencedirect.com/science/article/pii/S0886779817311306

Remittance and earthquake preparedness
https://www.sciencedirect.com/science/article/pii/S2212420915301898

**Organizations**

CASA People Helping People
www.casa-india.org/pages/index.php/who-we-are

Centre for Social Justice
www.centreforsocialjustice.org.uk

Partners of the Americas
www.partners.net

United Nations High Commissioner for Refugees UNHCR
www.unhcr.org

CS2 **Tools**

Emergence of Social Cohesion after a disaster (With reference to two flood affected locations in Colombo
https://www.sciencedirect.com/science/article/pii/S1877705818301371

Children and young people's wellbeing post-disaster: Safety and stability are critical
https://www.sciencedirect.com/science/article/pii/S2212420915300212

The reconstruction of transportation and environmental infrastructure in rural areas
https://www.sciencedirect.com/science/article/pii/S0386111212000192

A study on coordination between urban development and station capacity
https://www.sciencedirect.com/science/article/pii/S2213624X14000169

**Organizations**

Community Solutions
https://www.irex.org/project/community-solutions

Saferworld
http://www.saferworld.org.uk

International Research Institute for Climate and Society
http://portal.iri.columbia.edu/about

Center for Disaster Management and Risk Reduction Technology CEDIM
http://www.cedim.de/english/index.php/about

CS3 **Tools**

The role of communities in coping with natural disasters: Lessons from the 2010 Chile Earthquake and Tsunami
https://www.sciencedirect.com/science/article/pii/S1877705818301589

Social capital for disaster risk reduction and management with empirical evidences from Sundarbans of India
https://www.sciencedirect.com/science/article/pii/S2212420916301777

Healthcare and Disaster Supply Chain: Literature Review and Future Research
https://www.sciencedirect.com/science/article/pii/S2351978915011233

Mapping Construction Industry Roles to the Disaster Management Cycle
https://www.sciencedirect.com/science/article/pii/S2212567114009198

**Organizations**

International Organization for Migration (IOM)
http://www.iom.int/

The Office of U.S. Foreign Disaster Assistance (OFDA)
http://www.usaid.gov/our_work/humanitarian_assistance/disaster_assistance

Mercy Corps (MC)
http://www.mercycorps.org

Refugees International
http://www.refintl.org

**Figure 8.**
*A list of organizations and tools from the cluster text that belongs to the BMU closest to each query vector.*

supervision, ensures the success of the articulation. The decision-maker decides what kind of information is relevant for the decision-making based on particular concerns. In our case, as we focus mainly on architectural practices, the tools and organizations shown in the final outputs (**Figure 8**) are the ones with constructive and spatial focus.

The present experiment described a methodology that joined two discourses from the field of disaster response together to create a Common Ground, which then provided a ground for selecting and prioritizing information regarding a specific interest to articulate three final outputs. When working with a data-driven approach, there are often questions about whether the accuracy of the results can be trusted. To avoid this, the research proposed a methodology involving human and artificial intelligence interplay where accuracy is ensured by a series of filters that are user-dependent and secure the specificity of the result.

## 6. Conclusion and future work

First, several lists of humanitarian organizations indexed on the web contain information that cannot be compared; in other words, most organizations registered in one source are not present in another. To settle on a reliable source, one must navigate and filter through to get a representative number. Second, several approaches from different disciplines share similar keywords, e.g., academic writing from the field of health with the field of building safety assessments. Though both are extremely necessary, their approaches are entirely different. Therefore, even if AI predicts similarities among them, humans must be present to make the final decision and selection.

Additionally, it was found that there is a lack of research on how to integrate AI into a workflow of large-scale disaster response, especially in countries with scarce resources. Therefore, in the future, we should look for ways to apply the proposed or similar methodologies to an ongoing disaster case study to validate the speed and relevance of the results. Besides, it would be interesting if researchers add new discourses to those proposed in this experiment, e.g., social media, as these will bring a new stakeholder approach to disaster response. Also, researchers can examine different ways of encoding text data into numerical vectors. For example, instead of using word embedding, the frequency of words used over time (Google Books Ngram Viewer) can be used, and the results of the present research can be compared with the latter.

To conclude, AI is a problem-solving tool tailored to specific problems. Therefore, in a natural disaster scenario, this type of intelligence can be beneficial. Many problems gather a massive amount of data that require considerable computational power, and there are usually few people available to process this data. Therefore, by joining the strengths of human cognition with the strengths of AI computing, this experiment illustrates a method for creating a Common Ground where we can depict collaboration among humanitarian organizations and researchers around the world to aid an informed response in the aftermath of a natural disaster.

## Conflict of interest

The authors declare no conflict of interest.

## Notes/thanks/other declarations

To my family, Christian, Antonio, Clarita, Carlos, Carlos Jr. Luis.

## Author details

Karla Saldaña Ochoa
School of Architecture, The University of Florida, College of Design, Construction, and Planning, Gainesville, USA

*Address all correspondence to: ksaldanaochoa@ufl.edu

## References

[1] UNDACU. UN Disaster Assessment and Coordination (UNDAC) Field Handbook. 7th ed. 2018. ReliefWeb. Available from: https://reliefweb.int/report/world/un-disaster-assessment-and-coordination-undac-field-handbook-7th-edition-2018

[2] Imran M, Castillo C, Lucas J, Meier P, Vieweg S. AIDR Artificial intelligence for disaster response. In: Proceedings of the 23rd International Conference on World Wide Web. ACM. 2014. pp. 159-162

[3] Zhang D, Zhang Y, Li Q, Plummer T, Wang D. Crowdlearn: A crowd-ai hybrid system for deep learning-based damage assessment applications. In: 2019 IEEE 39th International Conference on Distributed Computing Systems (ICDCS). IEEE. 2019. pp. 1221-1232

[4] Shamoug A, Cranefield S, Dick G.. Information Retrieval for Humanitarian Crises via a Semantically Classified Word Embedding. 2018

[5] Fernandez-Luque L, Imran M. Humanitarian health computing using artificial intelligence and social media: A narrative literature review. International Journal of Medical Informatics. 2018;**114**:136-142

[6] Bünzli GM, Eppler F, Stewens T, Dinh BH, & JM. The Professionalization of Humanitarian Organizations The Art of Balancing Multiple Stakeholder Interests at the ICRC. (J. J. (Radboud U. Nijmegen), Ed.). . Springer International Publishing. 2019. DOI: 10.1007/978-3-030-03248-7 Library

[7] Tshitoyan V, Dagdelen J, Weston L, Dunn A, Rong Z, Kononova O, et al. Unsupervised word embeddings capture latent knowledge from materials science literature. Nature. 2019;**571**(7763):95-98

[8] Mikolov T, Chen K, Corrado G, Dean J. Efficient estimation of word representations in vector space. 2013a. arXiv preprint arXiv:1301.3781

[9] Mikolov T, Sutskever I, Chen K, Corrado GS, Dean J. Distributed representations of words and phrases and their compositionality. In: Advances in Neural Information Processing Systems. 2013b. pp. 3111-3119

[10] Li Q, Shah S, Liu X, Nourbakhsh A. Data sets: Word embeddings learned from tweets and general data. In: Proceedings of the 11th International Conference on Web and Social Media, ICWSM 2017. AAAI Press. 2017. pp. 428-436

[11] Socher R, Perelygin A, Wu J, Chuang J, Manning C, Ng A, et al. Recursive deep models for semantic compositionality over a sentiment treebank, EMNLP 2014. 2014

[12] DeCarlo LT. On the meaning and use of kurtosis. Psychological Methods. 1997;**2**(3):292

[13] Saldana Ochoa K, Ohlbrock PO, D'Acunto P, Moosavi V. Beyond typologies, beyond optimization: Exploring novel structural forms at the interface of human and machine intelligence. International Journal of Architectural Computing. 2021;**19**(3):466-490

[14] Hu X, Liu H. Text analytics in social media. In: Aggarwal C, Zhai C, editors. Mining Text Data. Springer: Boston, MA; 2012

[15] Ragini JR, Anand PMR, Bhaskar V. Big data analytics for disaster response and recovery through sentiment analysis. International

Journal of Information Management. 2018;**42**:13-24. DOI: 10.1016/j.ijinfomgt.2018.05.004

[16] Rytsarev IA, Kupriyanov AV, Kirsh DV, Liseckiy KS. Clustering of social media content with the use of BigData technology. Journal of Physics: Conference Series. 2018;**1096**(1):012085

[17] Mir I, Zaheer A. Verification of social impact theory claims in social media context. The Journal of Internet Banking and Commerce. 1970;**17**(1):1-15

[18] Pohl D, Bouchachia A, Hellwagner H. Automatic sub-event detection in emergency management using social media. In: Proceedings of the 21st International Conference on World Wide Web. ACM. 2012. pp. 683-686

[19] Lin RTK, Liang-Te CJ, Dai H-J, Day M-Y, Tsai RT-H, et al. Biological question answering with syntactic and semantic feature matching and an improved mean reciprocal ranking measurement. In: Proceedings of the 2008 IEEE International Conference on Information Reuse and Integration. 2008. pp. 184-189

[20] Kohonen T. Self-organized formation of topologically correct feature maps. Biological Cybernetics. 1982;**43**(1):59-69

[21] Moosavi V. Pre-specific modeling, [Doctoral Dissertation], Eidgenössische Technische Hochschule ETH Zürich, Nr. 22683. 2015. DOI: 10.3929/ethz-a-010544366

[22] Ecuador earthquake. 2016 Available from: https://www.worldvision.org/disaster-relief-news-stories/2016-ecuador-earthquake-facts. [Access: 08/22/2022]

[23] Central Valley of Costa Rica shakes with magnitude 5.4 quake. Available from: https://ticotimes.net/2019/11/24/central-valley-of-costa-rica-shakes-with-magnitude-5-4-quake. [Access: 08/22/2022]

[24] Strong Earthquake Strikes Indonesia, Killing at Least 20. Available from: https://www.nytimes.com/2019/09/25/world/asia/indonesia-earthquake-ambon.html. [Access: 08/22/2022]

Section 4

# Geotechnical Engineering Applications

Chapter 5

# Geophysical Investigations for Design Parameters Related to Geotechnical Engineering

*Folahan Peter Ibitoye*


## Abstract

Geophysical investigations for design parameters related to geotechnical engineering consider the application of geophysical methods such as electrical methods, seismic methods, magnetic prospecting method, electromagnetic prospecting methods, and these methods are needed to delineate the subsurface geological features before the erection of building structures, and design parameters used for geotechnical engineering can be investigated using geophysical methods. Moreover, geological structures such as fractures, faults, contacts, dykes, etc. can easily be delineated using geophysical methods, these geological structures are responsible for ground subsidence, building foundation problems, and building failures, which can affect the development of geotechnical engineering, and case studies will be referenced for easy understanding.


**Keywords:** engineering, geophysical, subsurface, fractures, design

## 1. Introduction

Geophysical investigations for design parameters related to geotechnical engineering provides a cost effective and faster way of civil engineering related investigations, the reliable results usually obtained from integrated investigations are indispensable in the pre-construction procedures, examples of such studies include [1–4]. In cases of large area of coverage for construction of engineering structures, it becomes necessary to adopt the use of geophysical methods as a means of reconnaissance procedure before embarking on geotechnical investigation, in order to increase safety net ratio, borehole drilling can be done without having environmental implication such as drilling a confined fracture which can lead to flooding, drilling of contaminant plumes prone subsoil which can cause groundwater contamination, omission of important subsoil information along distances between boreholes. In order to perform a sound geotechnical project design, the subsurface profile information must be obtained. subsurface exploration also known as geotechnical investigations usually involves drilling holes in the ground, retrieving soils or rock samples through the boreholes at predictable depths, extent of subsoil exploration depends on the spread size of the project (road, bridges, builds etc.). Geophysical investigations which are fast and cost effective avoid the destructive effects of drilling and can generate profile for the subsurface features. Bearing capacity is the capacity of the soil beneath foundation to support a super structure load. The maximum load-bearing capacity of the soil, that is the maximum stress then soil can carry without failure, For instance, in the basement complex of the south-western Nigeria, subsoil is categorized into sand, sandy-clay, clayey sand, clay, lateritic-clay, clay to sand ratio determines the bearing capacity of the soil beneath a foundation, the subsoil can be delineated by using geophysical investigations. when the ultimate bearing capacity of the soil beneath a foundation is exceeded by stress caused by the superstructure, the soil may compress and slide (shear) and a sliding(shear) surface may develop in the soil, this is called bearing capacity failure, this also manifests as cracks on the walls of a building [2]. Low subsurface bearing capacity results in a case where the foundation settles excessively, if the groundwater table is near the ground surface, it may affect the ultimate bearing capacity, but if the groundwater table is close to the ground surface, it may not affect the ultimate bearing capacity, groundwater table can be located through geophysical investigations.

When saturated soil is subjected to an external load, the pore water pressure increases immediately on the application of external load, with time, the increase of pore water pressure gradually decreases and effective stress gradually increases, as pore water drains from the soil, the pore volume and total volume of the soil gradually decrease, it is important to locate the geologic features such as faults, fractures, dykes using geophysical methods due to the fact that the drained water migrates from the soil through the geologic features which serve a conduits for movement of groundwater, but the soil exhibits weakness at these zones, therefore resulting into structural failure. The soil volume decrease in the vertical direction due to primary consolidation, thereby resulting into primary consolidation settlement, geophysical methods can ascertain the sand and clay compartments of subsoil that can result into differential settlement.

The essence of an exploration phase of geotechnical engineering is to identify the significant features of a typical geologic environment that may have significant impact on the proposed construction of an engineering structure. This includes the definition of the lateral distribution and thickness of the soil and rock strata within the zone of influence of the proposed construction; definition of groundwater conditions considering seasonal changes; identification of geologic hazards such as unstable slopes, faults, ground subsidence; identification of geologic materials for identification, classification and measurement of engineering properties

In this chapter, the applications of geophysical methods to derive the geotechnical parameters needed for building design are thoroughly explained. Section 2 discusses geotechnical parameters that can be derived from seismic refraction method. Section 3 discusses the application of the electrical method to the delineation of geotechnical parameters necessary for building construction, important factors such as subsurface layers and geologic features were explained. Section 4 illustrated the brief concept of magnetic method. Section 5 explains the application of Very Low Frequency Electromagnetic Method (VLF-EM). Section 6 explains relevant case studies.

## 2. Seismic exploration

The origin of seismic methods dates back to the early 1900s when instrumentation was designed to detect wave signals propagating through the earth arising from earthquakes. These waves propagated outwards from the focus (source) of the earthquakes and were detected and recorded by instruments on the surface of the earth. The study and analysis of the recorded signals resulted in the resolution of the source/focus and the magnitude of the earthquake. More importantly the nature of the internal structure of the earth's subsurface was well known from more detailed analysis of the form of the recorded waves and their travelled ray trajectories. These records showed waves that had propagated deep into the earth and had been reflected and/or refracted back to the surface from seismic/acoustic interfaces of the subsurface.

By the extension of these earthquake studies, the techniques of refraction and reflection seismology using artificial seismic sources were carried out about 1915 by Minthrop.

Seismic surveying has since been and still is the single most utilised geophysical surveying method in the search for oil and gas and also in hydrogeological, environmental and geotechnical problems. Seismic waves are generated and they propagate through the earth, get detected and recorded usually on the surface of the earth.

### 2.1 Seismic waves

Seismic waves are elastic waves that travel within the Earth; i.e. they spread out from a source by elastic deformation of the rocks through which they travel. This propagation depends on elastic properties that are described by the relationships between stress and strain. The linear relationship between stress and strain in the elastic range is specified for any material by its various elastic moduli), each of which expresses the ratio of a particular type of stress to the resultant strain. Seismic wave velocities are determined by the type of seismic wave and by elastic modulus and the density of the rocks they travel through. There are two groups of seismic waves: body waves and surface waves.

Body Waves: an elastic medium can be subjected to two types of deformation; namely the compressional/dilatational and Shear. Hence all the elastic waves are basically 'compressional/dilatational' or 'shear' waves. The essential difference between the two types is that one entails a volume change without rotation of the medium particles, whereas the entails rotation without any change of volume. These first two waves propagate along the surface or into the subsurface, returning to the surface by reflection or refraction.

- Compressional or P-(primary) wave: They cause a back-and-forth (compressional) motion which is parallel to the direction in which the wave is travelling.

- Shear or S waves: They cause a to-and-fro (shear) motion which is perpendicular to the direction in which the wave is travelling. Particle motions involve oscillation about a fixed-point at right angles to the direction of wave propagation. If all the particle oscillations are in the same plane, the shear wave is said to be plane-polarised. Unlike compressional waves, shear waves cannot travel through liquids or gases.

The equations of motion for dilatation(P-wave) and shear (S-wave) disturbances can be derived in terms of dilatational and rotational strain the results obtained from these equations is that the velocities of P- and S-waves ($V_P$ and $V_s$) respectively are related to the elastic moduli and density of material. The relationships are shown below:

### 2.2 P-wave velocity

$$Vp = \sqrt{\frac{K + \left(\frac{4}{3}\right)\mu}{\rho}} \tag{1}$$

This involves change of shape and volume

### 2.3 S-wave velocity

$$Vs = \left(\frac{\mu}{\rho}\right)^{\frac{1}{2}} \tag{2}$$

$$Vs = \sqrt{\frac{\varepsilon}{\rho}\frac{1}{2(1+\sigma)}} \tag{3}$$

This involves Change in shape only
Where,
$K$ = Bulk modulus
$\sigma$ = Poisson ratio
$\varepsilon$= Young modulus
$\mu$= Shear modulus/Lame's constant
$\rho$ = Density of the medium
These symbols are described in the diagram shown below:

Once the seismic wave velocities are measured, shear modulus ($\mu$), Bulk modulus ($K$), Young's modulus or modulus of elasticity ($\varepsilon$), Poisson's ratio ($\sigma$), Oedometric modulus ($\varepsilon_c$) and other elastic parameters may be obtained from the Eqs. (4)–(11) below. These expressions make the determination of the geotechnical parameters needed for building design as derived below:

1. Shear Modulus: Shear modulus ($\mu$) relates Shear wave velocity with acceleration due to gravity as expressed in Eq. (4)

$$\mu = \frac{\gamma V_s^2}{g} \tag{4}$$

Where $g$ is the acceleration due to gravity (9.8 m/s$^2$), where $g$ is given as $\gamma/_\rho$ , $\gamma$ is the unit weight of the soil and $\rho$ is the mass density. The unit mass density relates with P-wave velocity $V_p$ as shown in Eq. (5)

$$\gamma = \gamma_0 + 0.002V_P \tag{5}$$

$\gamma_0$ *is defined* as the reference unit weight value in KN/m$^3$ [3, 5, 6]. $\gamma o$ = 16 for loose, sandy and clayey soil. According to [5], some elastic parameters were defined in Eqs. (6)–(9):

2. Young's modulus/modulus of elasticity ($E$)

$$E = 2\mu(1+\sigma) \tag{6}$$

$\mu$ is shear modulus and $\sigma$ is the Poisson's ratio.

3. Oedometric modulus (*Ec*) given by Eq. (7)

$$E_c = \frac{(1-\sigma)E}{(1+\sigma)(1-2\sigma)} \tag{7}$$

$E$ is modulus of elasticity

4. Bulk modulus ($K$) is expressed by Eq. (8) as

$$K = \frac{2\mu(1+\sigma)}{3(1-2\sigma)} \tag{8}$$

5. Poisson's ratio ($\sigma$) is given as in Eq. (9) as

$$\sigma = \frac{\alpha - 2}{2(\alpha - 1)} \tag{9}$$

Where

$$\alpha = \frac{E_c}{\mu} = \left(\frac{V_p}{V_S}\right)^2 \tag{10}$$

6. Subgrade Coefficient (*Ks*), ultimate bearing capacity $q_f$ and allowable bearing pressure are given by Eqs (11)–(13) according as,

$$K_S = 4\gamma V_S \tag{11}$$

7. Ultimate Failure (Ultimate Bearing Capacity ($q_f$))

$$q_f = \frac{K_S}{40} \tag{12}$$

which is for shallow foundation.

8. Allowable Bearing Pressure ($q_a$)

$$q_a = \frac{q_f}{n} \tag{13}$$

Where $n$ is the factor of safety ($n$ = 4.0 for soils)

The basic requirement for construction or foundation sites is low compressibility and compliance and high bearing capacity which can be estimated from the reciprocal values of bulk modulus ($K$) and Young's modulus ($E$) respectively. Shear modulus and shear wave velocity of the soil layer is reduced with increasing shear strain [7]

## 3. Electrical method

The purpose of electrical resistivity survey is to determine the subsurface resistivity distribution by making measurements on the ground surface. Electrical resistivity method involves the passage of direct current, (DC) into the ground through two current electrodes ($C_1$ and $C_2$) while the resulting potential difference is measured across another pair of electrodes called potentials electrodes ($P_1$ and $P_2$), which may or may not be located within the current electrode pair, depending on the electrode array.

The apparent resistivity of the ground is calculated from the measured resistance (R). The survey techniques includes the Vertical Electrical Sounding (VES) and horizontal profiling (HP). Variations of the apparent resistivity with depth are measured in the VES technique, while lateral variations in ground resistivity are measured in the HP technique.

These variations in the resistivity of rocks are influenced by factors such as porosity, degree of fluid saturation, temperature, rock texture, rock types, geological processes and permeability. Geologic features such as fractures, fault zones, contacts can be easily delineated using the electrical, subsurface layers configuration can also be easily ranked based on their competence, soil corosivity can also be ranked and classified for burying of metallic structures during building constructions.

Outlined below are important areas of applications of electrical resistivity method in site characterization.

## 4. Depth to Bedrock determination

The determination of the overburden thickness and hence depth to the bedrock at a construction site or along the highway road is one of the major applications of electrical resistivity in site investigation. The depth to the competent bedrock is given by the total overburden thickness resting on the bedrock. Depth to bedrock is obtained from the summation of the thicknesses of the layers that constitutes the lithologic sequence in an area.

## 5. Structural mapping of the Bedrock

The structural setting of the bedrock (possible fault location, fracture, joints, buried river channels etc.) can be investigated with the seismic and electrical resistivity method. Fractures in bedrock occur most often in competent rocks unable to adjust to the stresses placed upon them. Fractures in bedrock are characterized by moisture making them more electrically conductive than a non-fractured bedrock. Fractured region may be topographically more depressed than the surrounding unfractured bedrock.

## 6. Location of construction materials

Electrical resistivity profiling method is mainly used in the search for sand and gravel deposits needed for construction projects. The data can be inverted to depict the resistivity variation, both laterally and vertically, against depth. The low-resistivity area (less than 20 ohm-m) corresponds to clay layers and high-resistivity zones (greater than 100 ohm-m) correspond to sand and gravel lenses.

## 7. Soil corrosivity

Corrosivity is defined as soil's ability to corrode a material that may be buried in it. As soils at building site would normally host metallic pipes etc., it becomes mandatory that a well-organized site testing exercise be carried out to evaluate soil aggressivity (corrosivity) taking into consideration the type of materials to be buried in it. The degree of corrosivity of the soil can be predicted using the electrical resistivity values. The electrical resistivity values of the top soil at the site can be used to assess the corrosivity of the soil of that site. Soils with resistivity values of less or equal to 10 ohm-m are strongly corrosive. Soils with resistivity values ranging between 10 and 60 ohm-m are moderately corrosive while those with resistivity between 60 and 180 ohm-m are slightly corrosive. Soils with resistivity values greater than 180 ohm-m are practically non corrosive.

## 8. Site subsoil competence

The strength of any geological material is influenced by several factors such as the mineralogy of its particles, the character of the particle contrast and agents of weathering. However, in a given locality, apparent resistivity values can be used for the evaluation of earth materials and their competence. Materials underlying a site can be judged to be generally competent or incompetent. High apparent resistivity zones are said to be competent in comparison with regions of relatively low resistivity values. Certain ranges of apparent resistivity values can be correlated with lithologic competence.

## 9. Mapping of seepage zones

Earth and rock-fill dams are large civil engineering structures designed to impound surface water. The design of dams makes provisions for control seepage and spillage. In spite of advances made in the field of geotechnical engineering, it is not possible to have 100% leak-proof structure. Anomalous seepages may sometimes occur through permeable soils, rock aquifers controlled by their structural bedrock topography and fault/joints structures. Geophysical method play an important role in mapping seepage paths and monitoring the changes of the seepage with time, enabling to plan technically and economically worthwhile remedial measures

## 10. Magnetic prospecting method

The origin of the earth's magnetism is commonly believed to be the liquid part of the earth's core, which cools at the outside as a result of which material becomes denser and sinks towards the inside of the outer core and new warm liquid matter rises to the outside; thus, convection currents are generated by liquid metallic matter which move through a weak cosmic magnetic field which subsequently generates induction currents. It is this induction current that generate the earth's crust magnetic field. Most rocks of the earth's crust contain crystals with magnetic minerals; thus most rocks have a certain amount of magnetism, which usually has two components induced by the magnetic field present while taking measurement, and remanent which is formed during geologic history [8]. The aim of magnetic survey is to investigate subsurface geology on the basis of anomalies in the Earth's magnetic field resulting from the magnetic properties of the rocks [9]. Although most rock-forming minerals are effectively nonmagnetic, certain rock types contain sufficient magnetic minerals to produce significant magnetic anomalies. An observed anomalous magnetization might be associated with buried magnetic objects that are potentially of commercial interest. Anomalies recorded in the measured field are interpreted in terms of variations in magnetic susceptibility and/or remanent magnetism, the physical rock properties affecting the measurements. Magnetic susceptibility is the physical property on which the response of magnetic method is based, and it is the property whose distribution we are trying to investigate.

## 11. Very low frequency electromagnetic method

Very Low Frequency (VLF-EM) is an effective reconnaissance geophysical tool for mapping geologic features. It may be used wherever an electrical conductivity contrast exists between geological units, this includes Fault mapping, Groundwater investigation, Overburden (summation of all the materials on the bedrock) mapping, contaminant mapping. Electrical conductive features include fault zones which tend to be more conductive than the surrounding bedrock or host rock. Other conductive geologic units include moist, clayey or fine grained soils which tends to be more conductive than dry, sandy or coarse grained soils. Hence, these geologic objectives are reasonable "targets" and can be mapped using electromagnetic methods. In reconnaissance mode, VLF profiles can be run quickly and inexpensively to determine anomalous areas which requires further investigation.

## 12. Case studies: application of electrical and very low frequency electromagnetic method to pre-construction investigations

### 12.1 Site descriptions, geomorphology and physiography

The study area as shown in **Figure 1** is located within Igarra, Southwestern Nigeria expressed in Universal Traverse Mercator (UTM) Zone 3, and the study area is located within Northings 807054 mN and 807265 mN and Eastings 179000 mE and 179210 mE. The landforms are of two main groups, the high hills which are found in the north-eastern part of Igarra, while the other is low lying around north-western and south-western part of Igarra. The Igarra area is made up of untarred roads, footpaths, rivers, the tributaries converges at the point to form rivers which flows along the strike direction of the outcrops in the area.

### 12.2 Brief geology of the study area

Three major rocks overlie the basement complex in the Igarra area. The successions of the rocks are namely;

- Granodiorite which are rich in biotite and hornblende, diorite which are unmetamorphosised pegmatite which has large quartz clasts.

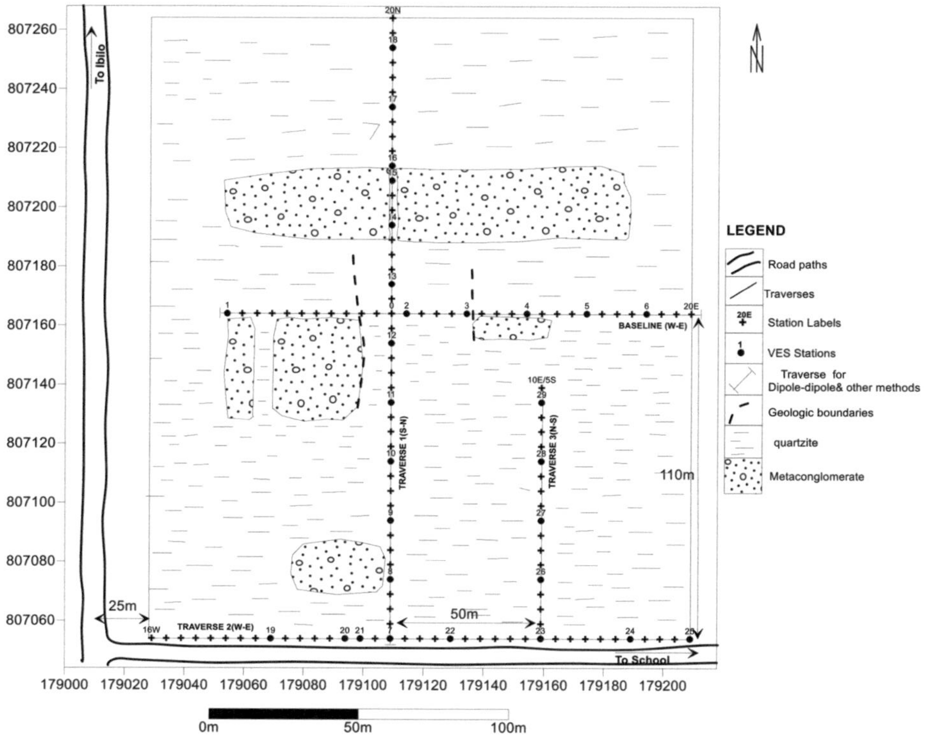


**Figure 1.**
*Data acquisition map of the study area.*

- Rocks formed under low temperature and pressure(low grade metasediments) such as schists, calc-silicate gneiss, marble, metaconglomerate and quartzites.

- Metamorphic rocks such as gneiss rich in biotite-hornblende with bands of minerals, the metasediments are younger than the basement complex [10] and the older granites represents the youngest group of rocks of the Precambrian age in this area.

The sequence of rocks in the Igarra area is made up of four main groups which includes Calc-silicate gneiss and marble, Metaconglomerate, Mica-schist and Quartz-biotite schist

## 12.3 Method of study

The geophysical investigation involved electrical resistivity method (Dipole-Dipole Technique) and Very Low Frequency-Electromagnetic Method, the steps includes geologic mapping (reconnaissance survey of the study area) and establishing of traverses, production of geophysical location maps from the base/topographical map, data acquisition for all the geophysical methods followed by processing and interpretation. Inverted 2D Model of VLF data was generated using K-H Filter (Karous-Hjelt Filtering software). The Dipole-Dipole array was used for the data acquisition. The inter-electrode spacing (a) of 5 m was adopted while inter-dipole separation factor (n) was varied from 1–5. The apparent resistivity values were calculated using $\pi n(n + 1)(n + 2)a$ as the geometric factor. 2-D inversion modelling of the Dipole-Dipole data was carried out using DIPRO$^{TM}$ Software developed by the Korea Institute of Geoscience and Mineral Resources

## 12.4 2-D resistivity structure along South-North direction

The dipole-dipole pseudosections and the 2D resistivity structure along the S-N direction are shown in **Figure 2**. The 2D resistivity structure revealed four geoelectric/geologic subsurface layers marked A, B, C and D separated by geologic boundaries; namely topsoil marked by A (generally blue colour except at few points with green colour), B (green and blue colour), weathered/fractured basement marked by C (generally green except at few points with yellow colour).The topsoil is generally thin and subsume into the weathered layer in many places due to its thickness and it is characterised mainly by clay but few sandy clay. The weathered layer is characterised by clay and sandy clay, its thickness increases from the centre of the traverse towards the southern and the northern flanks respectively. The northern flank of the weathered layer exhibit very low resistivity values which may be due to high level of saturation and diagnostic effect of wet clay.

The weathered/fractured basement is characterised by two linear features which are between stations 4 and 5(distance at 40 m and 50 m) and stations 14 and 16

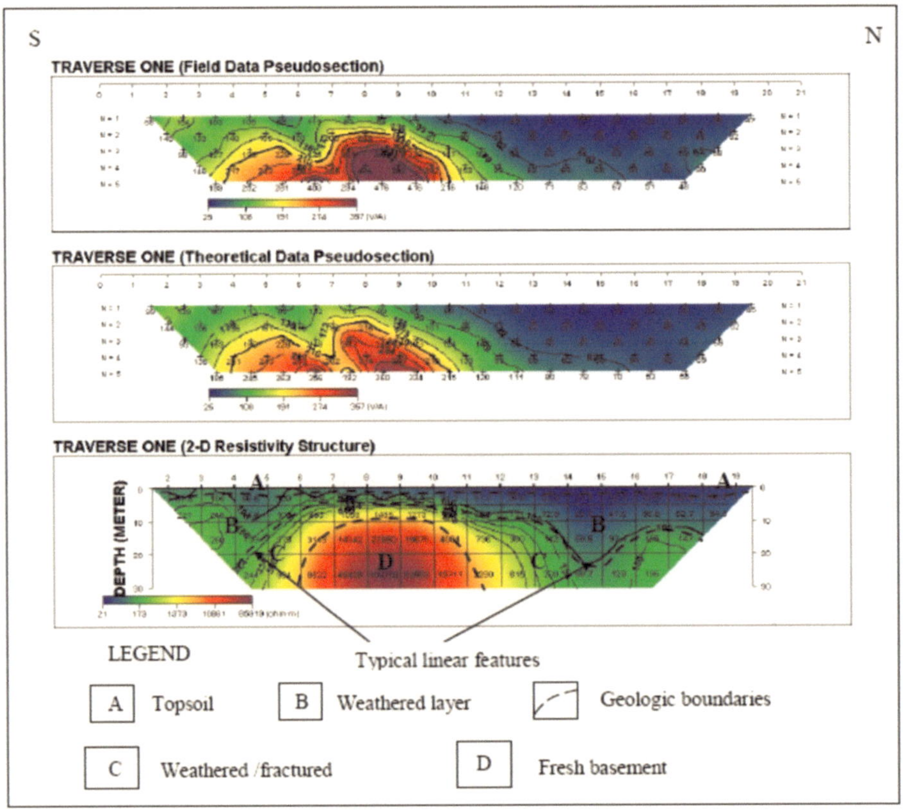


**Figure 2.**
*2-D modelling of dipole-dipole along S-N direction.*

DOI: http://dx.doi.org/10.5772/intechopen.108712

(distance at 140 m and 160 m). The upper part of the weathered/fractured basement exhibits lower resistivity than its lower section, signifying that the upper part (green colour) is more weathered than the lower part of the pseudosection.

The depth to bedrock is thinner between stations 7 and 10 (distance at 70 m and 100 m), but increases towards the southern flank (between stations 2 and 7) and increases towards the northern flank(stations 10 and 19), this shows that the bedrock extends beyond the depth of study (30 m) between the stations 2 and 6 and stations 12 and 19. The overburden at the northern flank is characterised by higher portion of clay due to its lower resistivity values when compared with the southern flank.

### 12.5 2-D resistivity structure along West-East direction

The 2D resistivity structure (**Figure 3**) reveals four geologic/geoelectric subsurface layers separated by geologic boundaries; namely topsoil marked by A (generally blue except few points with yellow and green colour), weathered layer marked B (generally blue except few points with yellow and green colour), weathered/fractured

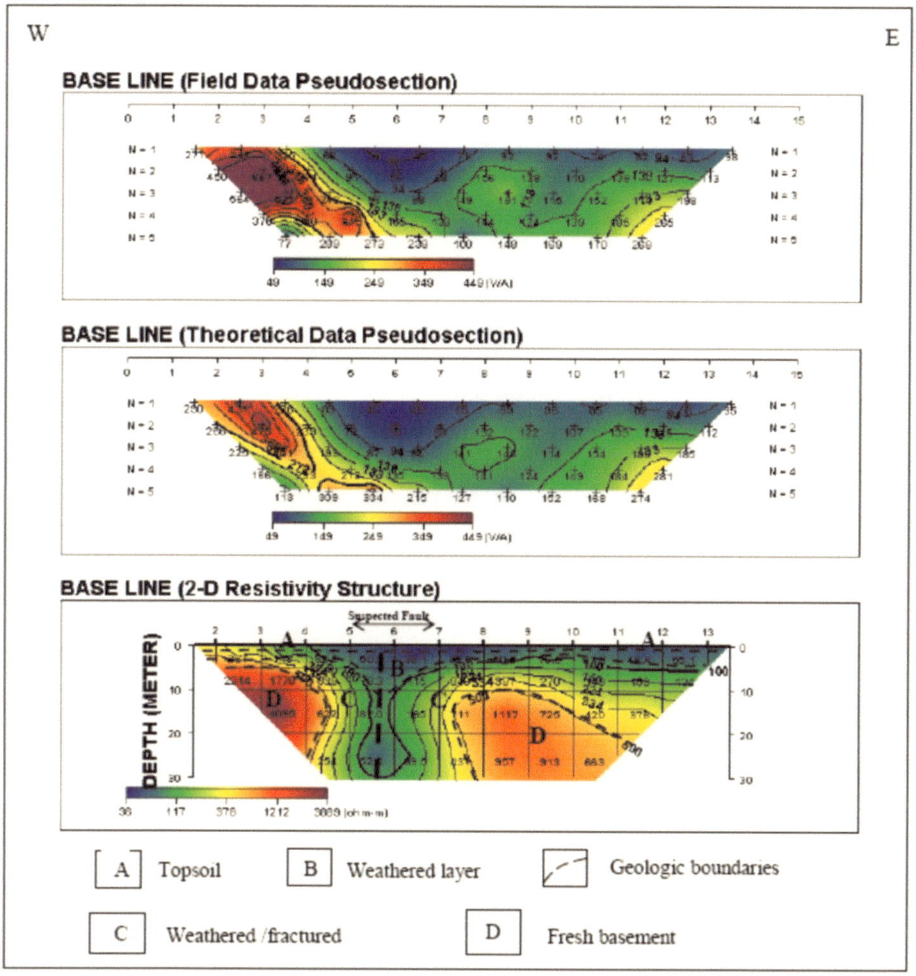


**Figure 3.**
*2-D modelling of dipole-dipole along West-East direction.*

basement marked C (green and yellow colour) and fresh basement marked D (generally red except few points with yellow colour). The topsoil is generally thin but subsume into the weathered layer in many places due to its thickness, the topsoil is characterised mainly by clay with few portions of laterite and sandy clay. The weathered layer is characterised mainly by clay (lowest resistivity represented by blue colour) between stations 5 and 8 (distance at 50 m and 80 m), but the resistivity values increases from the above stations towards the western and the eastern flank signifying the reduction of clay to sand ratio.

A linear feature is noticed between stations 5 and 7 (distance at 50 m and 70 m) which has significant depth extent. The low resistivity zone is between stations 5 and 7 (50 m–70 m) which could be diagnostic of a suspected fault zone with a width of approximately 19 m. The suspected fault zone is flanked on both sides by regions of higher resistivity. The extremely low resistivity value (blue colour) that was noticed at the lower part of the fault zone signifies the effect of high saturation and diagnostic effect of wet clay. The resistive parts are seen at the lower part of the section which is the fresh basement.

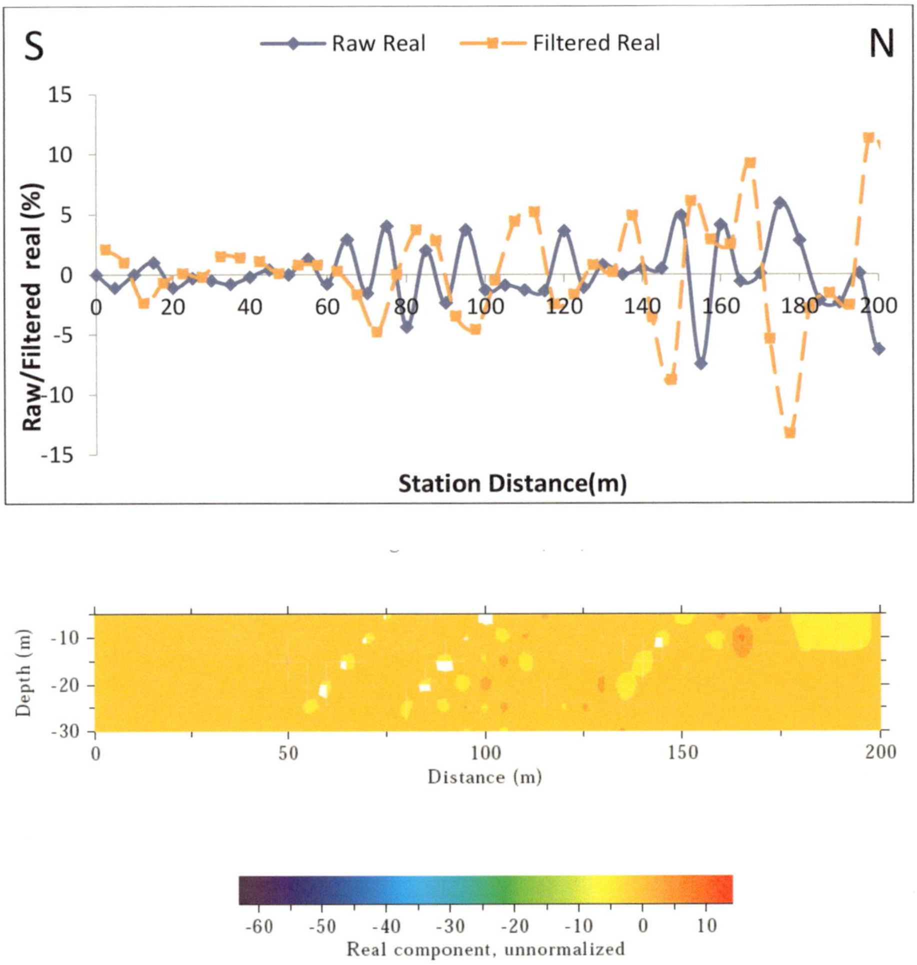


**Figure 4.**
*VLF profile and inverted 2D model obtained along traverse one (S-N).*

### 12.6 VLF profile along South-North and West-East direction

On the traverse one which is trending South-North, the VLF-EM profile (**Figure 4**) identified peak positive filtered real values at distances 2 m, 40 m, 83 m, 110 m, 138 m, 168 m and 200 m. The amplitude of the peak positive filtered real values are very low, the inverted model shows that most of the peaks manifest as non-anomalous conductive zones, except for the peak positive filtered real noticed at 200 m. The VLF-EM profile and the inverted model are shown on the profile.

On the traverse trending North-South, the VLF-EM profile (**Figure 5**) identified peak positive filtered real values at distances 18 m, 43 m and 58 m. The amplitude of the peak positive filtered real values are very low, the inverted model shows that most of the peaks manifest as non-anomalous conductive zones, except for the peak positive filtered real noticed at 18 m.

On the baseline which is trending West-East direction, the VLF-EM profile and the inverted model are shown in **Figure 6**, the VLF-EM profile identified peak positive filtered real values at distances 18 m, 32 m, 46 m, 60 m, 82 m, 112 m and 138 m, the

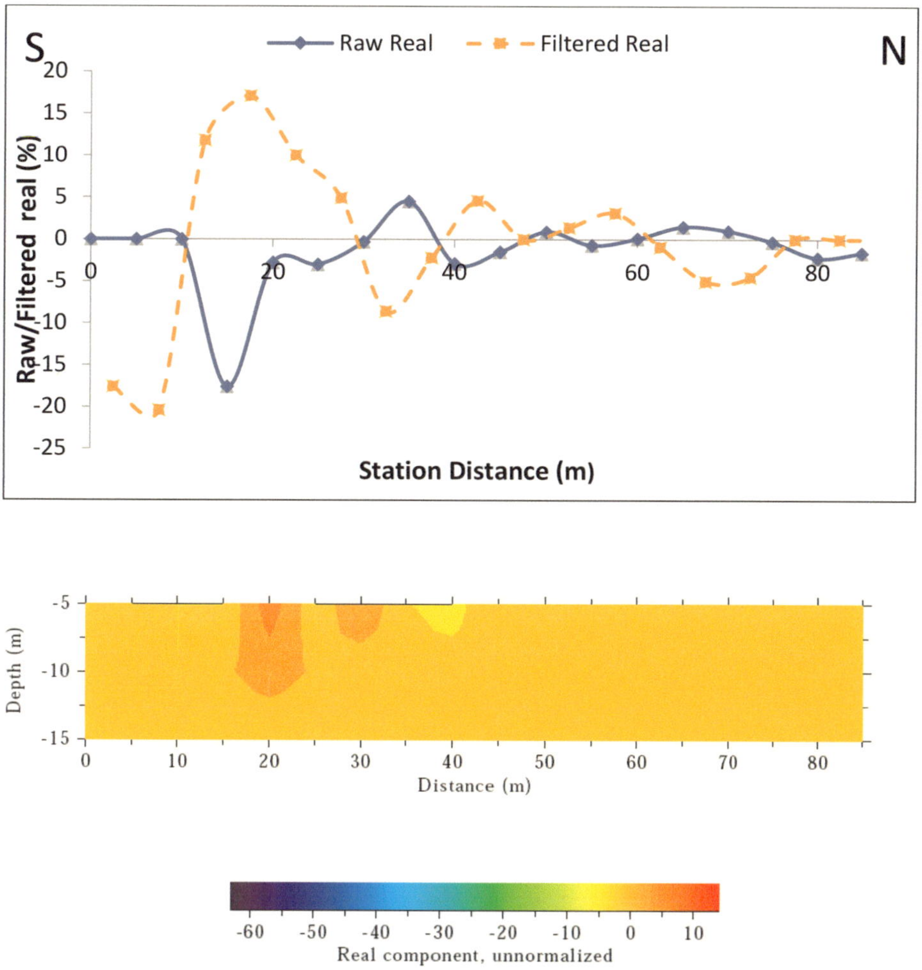


**Figure 5.**
*VLF profile and inverted 2D model obtained along traverse three (South-North).*

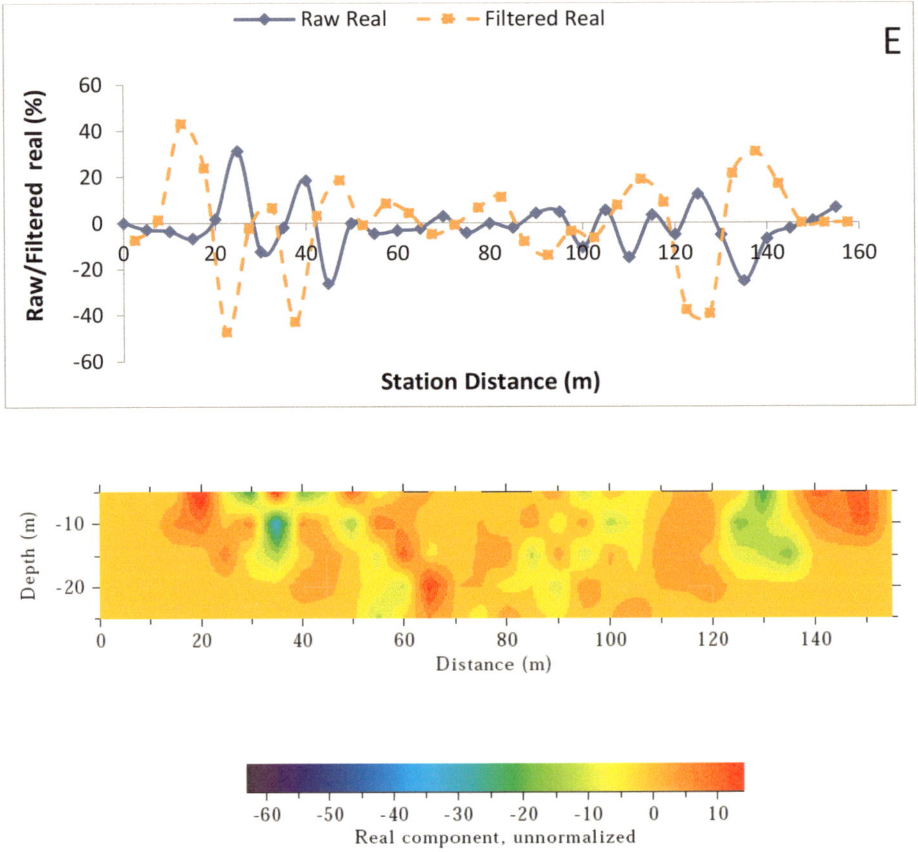


**Figure 6.**
*VLF profile and inverted 2D model obtained along baseline (West-East).*

observation agree with conductive zones delineated by the inverted model at distance 20–60 m, 80–100 m and 122–138 m. The conductive zone between distance 122–138 m and 20–60 m are typical of a linear feature (fracture) because of its depth and it is dipping to the west.

On the traverse which is trending West-East, the VLF-EM profile and the inverted model are shown in **Figure** 7. The VLF-EM profile identified peak positive filtered real values at distance 78 m, 102 m, 118 m and 138 m, the observations agree with the conductive zones delineated at distances 3–20 m, 110–140 m, the conductive zones between 110 m and 140 m is typical of a linear feature (fracture) and it is dipping to the east.

## 13. Conclusion

In this chapter geophysical investigations for design parameters related to geotechnical engineering was explored.

The strength of the geophysical methods lies in the ability of the method to image the subsurface in a faster and cost effective approach when compared with geotechnical investigations. Geophysical methods provides adequate guidance to civil engineers before embarking in boring and drilling operations, this will avoid excessive

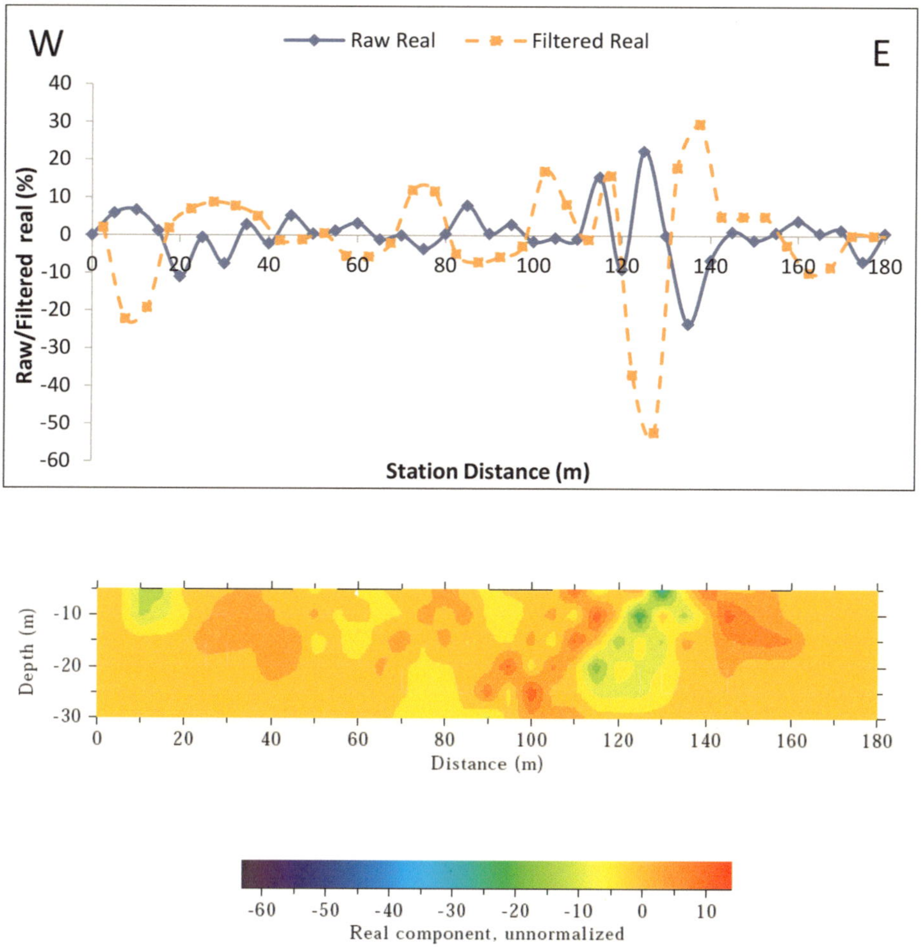


**Figure 7.**
*VLF profile and inverted 2D model obtained along traverse two (West-East).*

damage of subsurface materials which result in environmental risks such as flooding activities.

Seismic refraction, electrical methods, electromagnetic method and magnetic method have been proven to be useful for evaluating design parameters/geologic subsurface deductions needed before the construction of engineering structures such as buildings, rail ways, dams, bridges etc. Case studies involving the application of electrical method and Very Low Frequency Electromagnetic method at Study area within Igarra, South-western Nigeria was explained to reveal the importance of geophysical methods in geotechnical engineering. Geologic features such as faults and fractures were delineated from the Electrical Resistivity Image of the subsurface, VLF profile and the Inverted 2D Model, the compositions (sandy clay, clay, clayey sand and laterite) of the subsurface/geoelectric layers were also classified based on their competence.

In conclusion, the importance of geophysical investigations for evaluation of the design parameters related to geotechnical engineering cannot be neglected in the pre-construction phase.

## Author details

Folahan Peter Ibitoye
Prototype Engineering Development Institute (National Agency for Science and Engineering Infrastructure), Ilesa, Nigeria

*Address all correspondence to: folaibitoye@yahoo.com

## References

[1] Akintorinwa OJ, Adesoji JI. Application of geophysical and geotechnical investigations in engineering site evaluation. International Journal of Physical Sciences. 2009;**4**(8):443-454

[2] Ibitoye FP, Ipinmoroti FV, Salami M, Akinluwade KJ, Taiwo AT, Adetunji AR. Application of geophysical methods to building foundation studies. International Journal of Geosciences. 2013;**4**(9):1256-1266

[3] Adewoyin OO, Joshua EO, Akinwunmi II, Omeje M, Joel ES. Evaluation of geotechnical parameters using geophysical data. Journal of Engineering Technology and Science. 2017;**49**(1):95-113

[4] Teshome AM. Application of magnetic method to characterize Abaya Campus building site, Southern Ethiopia. Journal of Geology & Geophysics. 2022; **11**(1):1-5

[5] Tezcan SS, Ozdemir Z, Keceli A. Allowable bearing capacity of shallow foundations based on shear wave velocity. Journal of Geotechnical and Geoenvironmental Engineering. 2006;**24**: 203-218

[6] Atat JG, Akpabio IO, George NG. Allowable bearing capacity for shallow foundation in Eket local government area, Akwa Ibom State, Southern Nigeria. International Journal of Geosciences. 2013;**4**:1491-1500

[7] Massarsch KR. Deformation properties of fine-grained soils by means of seismic tests. In: International Conference on Site Characterization, ISC'2. Porto; 2004. pp. 133-146

[8] Telford WM, Geldart L, Sheriff RE. Applied Geophysics. Second ed. Cambridge University Press; 1990. pp. 63-75

[9] Kearey P, Brooks M. An Introduction to Geophysical Exploration. third ed. Blackwell Publishing; 2002. pp. 155-198

[10] Odeyemi IB. Preliminary report on the field relationships of the basement complex rocks around Igarra, Midwest Nigeria. In: Kogbe CA, editor. Geology of Nigeria. Lagos: Elizabethan Publishing Co; 1976. pp. 59-63

Chapter 6

# Rock Slope Stability under Temperature Fluctuations

*Dagan Bakun-Mazor*


## Abstract

Air temperature fluctuations cause intermittent shrinkage and expansion of rock blocks close to the surface. These cyclic deformations can bring cumulative and irreversible displacement of blocks down rock slopes, creating potentially dangerous conditions. This chapter presents examples from stations monitoring various slopes and natural cliffs that illustrate this phenomenon. The chapter will also present the results of a laboratory experiment performed on a block system that simulates the typical geological structure of a slope in a stratified and cracked rock mass. Inside a tensile crack, a prismatic rock block serves as a wedge capable of accelerating the cumulative displacement. The results obtained from the laboratory measurements serve as a basis for calibrating analytical and numerical 3D models of the problem. The results of the laboratory experiment and the numerical models clearly show that the wedge is of great importance in accelerating the rate of movement of the blocks on the slope. Further numerical analysis performed on blocks on the slopes of Mount Masada shows that temperature changes alone could explain the blocks' detachment from the slope.


**Keywords:** rock slope stability, wedging mechanism, thermomechanical response, 3DEC, field monitoring observation

## 1. Introduction

Observations from monitoring stations in rock slopes around the world indicate thermally induced cumulative deformations of removable rock blocks. Although temperature fluctuations are cyclical, the reaction of the rock is not necessarily so, indicating that the mere accumulation of plastic deformations that weaken the mass of the rock can lead to rock failure.

The first time the research group of which I was a member noticed the phenomenon was when monitoring blocks that had detached from the rock mass in the eastern cliff of Mount Masada, in Israel's Judean Desert. During 1998, as part of a project to build a new cable car to the mountaintop, the group monitored blocks whose stability was questionable [1]. Joint meters were installed to measure joint displacement along the tensile joints that separate the blocks from the rock mass. Temperature and humidity meters were also installed.

The location of Mount Masada along the Dead Sea transform, just near the Dead Sea, is shown in **Figure 1A**. The main monitored block (referred to as "Block 1") is shown in the picture in **Figure 1B**, taken before the construction of the new station.

(The old cable car station is shown beneath the block.) In the picture in **Figure 1C**, taken after the completion of the new station and the access road from it to the mountaintop, you can see the proximity of the block to the visitor area. The results of the monitoring station are shown in **Figure 1D**. The results indicate a direct relationship between seasonal temperature changes and joint displacement [1]. At the end of that monitoring campaign, it was decided to anchor the block to the rock mass.

Another observation from Masada was obtained from a research station installed on the western side of the mountain in 2009 [2]. This station was installed near the Roman Ramp (**Figure 2A**). The station consisted of four joint meters, an air temperature sensor, and a humidity sensor (**Figure 2B**). The monitoring results over 2 years are shown in **Figure 2C**, where joint openings are shown in black lines, and air temperature is in blue. A cumulative opening of the joints can be seen as a function of air temperature.

Various other studies by worldwide monitoring stations noticed the cumulative displacement of blocks on rock slopes as a response to temperature fluctuations. For example, Vlcko et al. [3] monitored selected castles on rock cliffs to preserve cultural

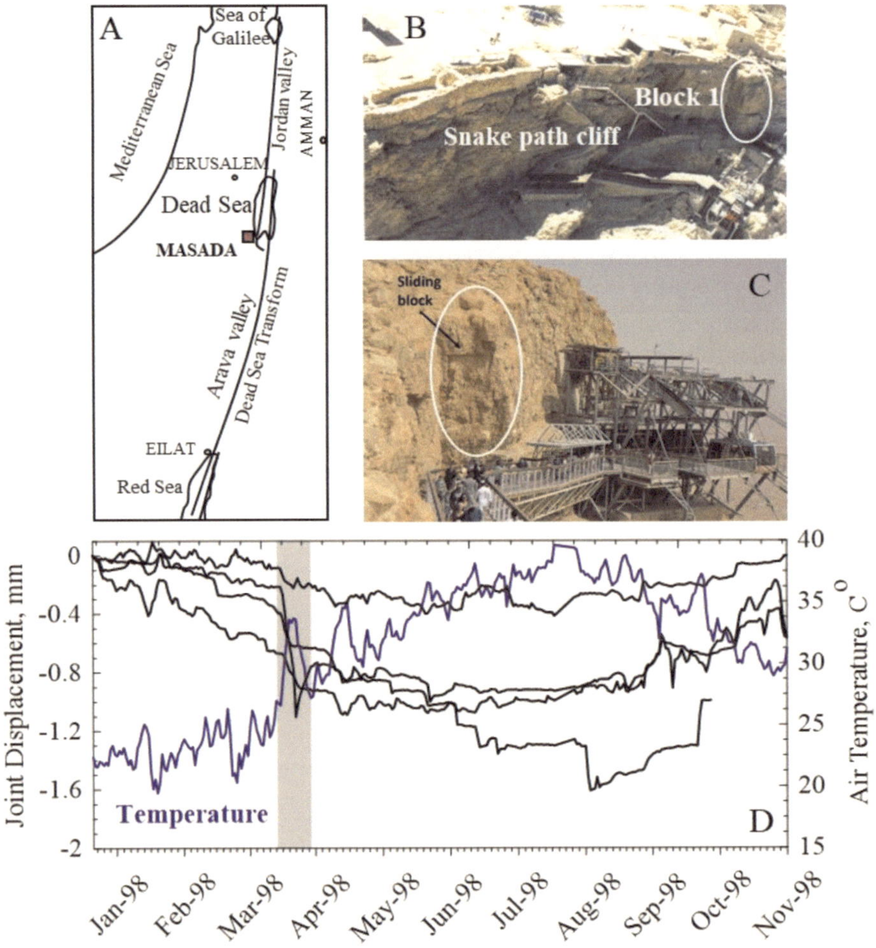


**Figure 1.**
*Observation from a monitoring station at the eastern cliff of Masada. A) Location map, B) the snake path cliff with the old cable car station, C) the monitored sliding block ("Block 1"), and D) the obtained monitoring results during 1998 (After Hatzor et al. [1]).*

**Figure 2.**
*Observation from the monitoring station at the western cliff of Masada. A) Station location, B) station setup, and C) obtained results (After Bakun-Mazor et al. [2]).*

heritage sites. They reported a thermal response of rocks as a result of seasonal periodic temperature changes.

Gischig et al. [4] monitored the rock slope instability above Randa (Switzerland) and measured temperature-controlled deformation trends at depths up to 68 meters. They interpreted this seasonal deformation trend as being controlled by thermomechanical effects driven by near-surface temperature cycles. Natural rock slope deformations across fractures, predominantly in a chert rock mass, were monitored by Mufundirwa et al. [5]. They proposed a new method to minimize displacement proportional to temperature and reported thermally induced rock mass movements. In their pioneering study, Gunzburger et al. [6] suggested that temperature fluctuations cause cumulative deformations in systems with a preferred sliding surface directionality.

These are just a few works from a wide variety of projects for monitoring rock movements along slopes. One of the difficulties is isolating the temperature's effect from the other factors that can lead to slope failure, as discussed by Fiorucci et al. [7]. Moreover, asymmetry in the block system leads to the development of preferred directional stresses that can induce permanent displacement of rock blocks.

This chapter demonstrates a mechanism capable of explaining the cumulative displacement of rock blocks on a slope due to temperature fluctuations. First, I present the mechanism and describe laboratory procedures that tested the mechanism and, and then the use of numerical simulations to analyze a case study in the field.

## 2. The Ratchet mechanism

### 2.1 Mechanism description

The ratchet model is one of the mechanisms that can explain cumulative displacement driven by temperature changes. A schematic diagram depicting the mechanism in a stratified and jointed rock mass is shown in **Figure 3**. The components of the mechanism are shown in **Figure 3A**. A sliding block detached from the rock mass at a tension crack lies on a sliding surface. Rock fragments inside the tension crack function as wedge blocks in the ratchet mechanism. During cooling episodes (**Figure 3B**), the system contracts, and the tension crack opens. As a result, the rock fragments slide into the gradual opening of the crack. During heating episodes (**Figure 3C**), the system expands. The sliding block also expands and aims to close the tension crack. However, the rock fragments inside the crack are locked in place, resulting in the sliding block moving down the slope.

A striking example of this can be inferred along a rock column in the Larzac Plateau (Southern France), as reported by Taboada et al. [8] and shown in **Figure 4**. It has nicely appeared that the rock column has a thermomechanical creep, and permanent deformations are associated with mechanical forces induced by short-term thermal cycles. Additional examples of potential wedging mechanisms are shown in the images in **Figure 5**.

A one-dimensional analytical solution to the mechanism mentioned above was developed by Pasten [9]. The solution, which also is presented in detail in Bakun-Mazor et al. [10], consists of three displacement components. One component depends on the thermal expansion of the block system. The other two components depend on the ability of the sliding block to absorb some of the deformations elastically through elastic deformation of both the sliding block and the sliding surface upon which it rests. However, if the thermal expansion component is greater than the elastic components, the block is expected to accumulate plastic deformations reflected in its relative sliding down the slope.

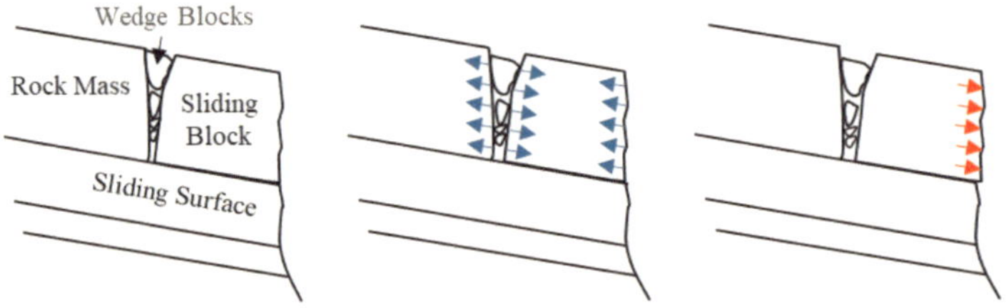


A B C

**Figure 3.**
*The ratchet model. A) Model components along a rock slope, B) rock contraction at cooling episode, and C) rock expansion at heating episode.*

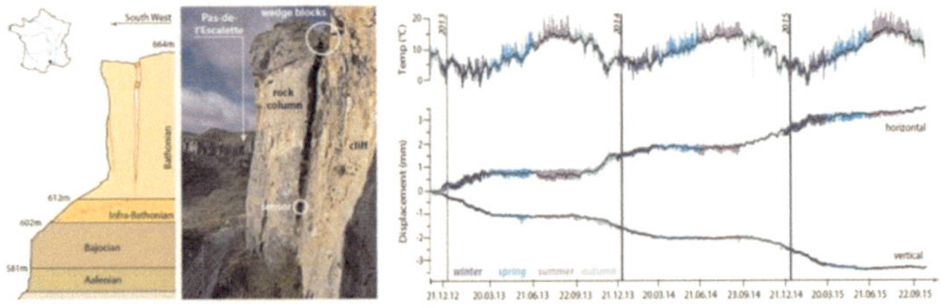


A B

**Figure 4.**
*Example from monitoring station at Larzac Plateau (Southern France), measuring thermally induced rock column displacement. A) The rock column with the wedge blocks, and B) temperature and displacement data from the rock column (After Taboada et al. [8]).*

**Figure 5.**
*Photos of blocks from the field that illustrate the ratchet mechanism, A) Masada Mountain, Israel, B) Arugot Valley, Israel, C) Yellow Mountain, China, D) Zohar Valley, Israel, and E) Machtesh Ramon, Israel. The scale of each bar in the inset represents 1 m.*

### 2.2 Lab experiments inside temperature-controlled chambers

Since the proposal of the mechanism, it has been tested by several researchers in the laboratory. The first to test the feasibility of measuring cumulative displacement as a result of a temperature change were Pasten et al. [11]. Their model, built from an acrylic block and wedge system, was placed in a temperature-controlled chamber that was heated and cooled using a light bulb and a fan. They showed that the wedge slides down due to temperature fluctuations and that the accumulation of plastic displacement, induced by temperature cycling, are proportional to the period and amplitude of the input temperature signal.

Another work that examined the mechanism in the laboratory was conducted by Greif et al. [12], who tested sandstone samples. As part of the work, the researchers examined how the ratio between the wedge length and the sliding block length affects the cumulative displacement. They then compared the measured results to those obtained from the analytical model. They reported that the results of the physical model agreed with Pasten's analytical solution [9].

### 2.3 Lab experiment on a large scale

The two studies mentioned above tested the mechanism on relatively small models inside thermal chambers in the laboratory. On the other hand, a research work I carried out with the research group examined the mechanism of a model representing the dimensions of blocks in the field. The model was prepared using concrete castings formed as a block assembly representing a typical situation in stratified and jointed rock slopes found in the field. **Figure 6A** shows the block system with its dimensions in cm. The system model was placed on a tilted steel table and placed inside a temperature-controlled room (**Figure 6B**).

**Figure 6** shows the four locations where displacement transducers were installed to measure: A) the relative displacement between the sliding block and the fixed block, B) the vertical displacement of the wedge, C) the displacement of the front of the sliding block, and D) the thermal response of the concrete block (this transducer used as a dummy).

The relative displacement of the block assembly was measured using both displacement transducers and a high-resolution visual range camera (**Figure 7**). The visual range camera tracked the junction area between the four blocks, as shown in

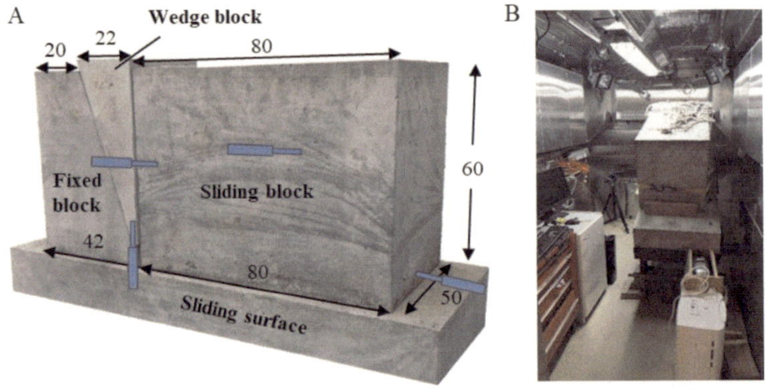


**Figure 6.**
*The physical model in the lab. A) Concrete block system with the location of the displacement transducers (dimensions in cm), and B) the block system on an inclined steel table inside a temperature-controlled room.*

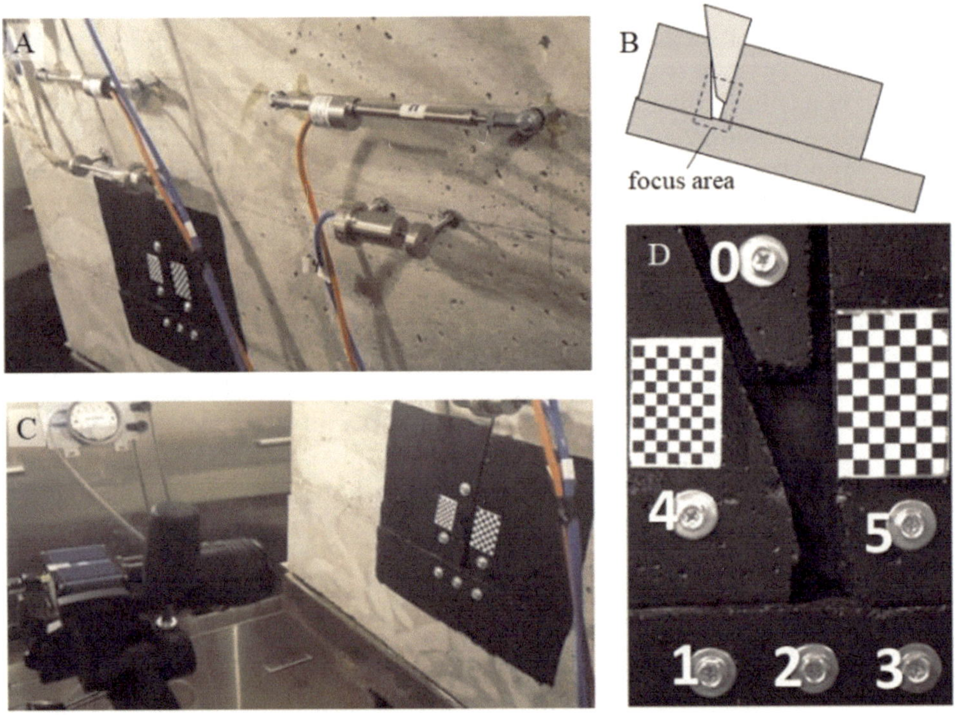


**Figure 7.**
*The experiment measurement setup. A) Displacement transducers, B) the focus area along the junction between the blocks, C) visual range camera tracing the focus area, and D) markers on the blocks for image processing.*

**Figure 7**. Screws were installed on the blocks to serve as markers for image processing (**Figure 7D**). The material properties of the concrete were measured in the laboratory and are detailed in **Table 1**. The room temperature is heated using electrical furnaces and cooled using an air conditioning system that operates inside air sleeves to prevent air movement inside the room. A controller regulates the thermal system. In a preliminary phase of the laboratory experiment, we tested the time needed for the center of the sliding block to reach the target temperature of the room. The measurements

| | Parameter | Symbol | Units | Value |
|---|---|---|---|---|
| Material properties | Elastic modulus | $E$ | GPa | 24.5 |
| | Poisson's ratio | $\nu$ | - | 0.251 |
| | Bulk density | $\rho$ | kg/m$^3$ | 2140 |
| Joint properties | Friction angle | $\phi$ | ° | 21 |
| | Normal stiffness | $K_n$ | GPa/m | 5 |
| | Shear stiffness | $K_s$ | GPa/m | 0.5 |
| Thermal properties | Thermal expansion coefficient | $\alpha$ | $10^{-6}$/°C | 3.22 |
| | Thermal diffusivity | $D_T$ | $10^{-7}$m$^2$/sec | 5.44 |
| | Specific heat capacity (assumed) | $C_p$ | J/kg/K | 850 |

**Table 1.**
*Thermomechanical properties of the concrete used for the experiments.*

from the thermocouples inside the block showed that it takes about 3 days for the block to reach room temperature in its center. During the experiment, the room temperature changed intermittently from 35 degrees to 5 degrees, with a delay of about 75 hours between one target temperature and another.

The experiment's results indicate cumulative sliding of the block down the slope, depending on the temperature cycles. **Figure 8A** shows time-dependent cumulative displacement over three and a half temperature cycles. The block responds to the temperature cycles but accumulates a one-directional slip down the slope. The slip reaches about 0.06 mm at the end of three temperature cycles, that is, a rate of about 0.02 mm per cycle. **Figure 8B** shows the wedge sliding down to the gradual opening of the crack.

## 3. Numerical simulations

### 3.1 Validation against laboratory results

To deeply study the mechanism and carry out simulations of a case study in the field, we performed a numerical analysis using a three-dimensional version of the Distinct Element Code (3DEC) [13, 14]. The theoretical foundation of this method is the formulation and solution of equations of motion of deformable blocks by an explicit time-marching scheme using the finite volume method. The code can simulate the response of discontinuous media to static, dynamic, or thermal loading and can provide the corresponding deformation.

We first reproduced the geometry of the physical model in the 3DEC environment for validation. We applied the thermal boundary conditions on all exposed faces as the temperature-time histories recorded in the lab. The model was fixed in all directions at the bottom of the sliding surface, and in the normal direction behind the rock mass. The numerical simulation was performed with the thermomechanical parameters exactly as measured in the laboratory. The response of the block system in the numerical model is shown in **Figure 9**. The block's response in the numerical analysis appears slightly greater than the displacement measured in the experiment. The numerical model, however, captures the expected physics of the failure mode very well. We performed sensitivity analyses to better explore the relative influence of the

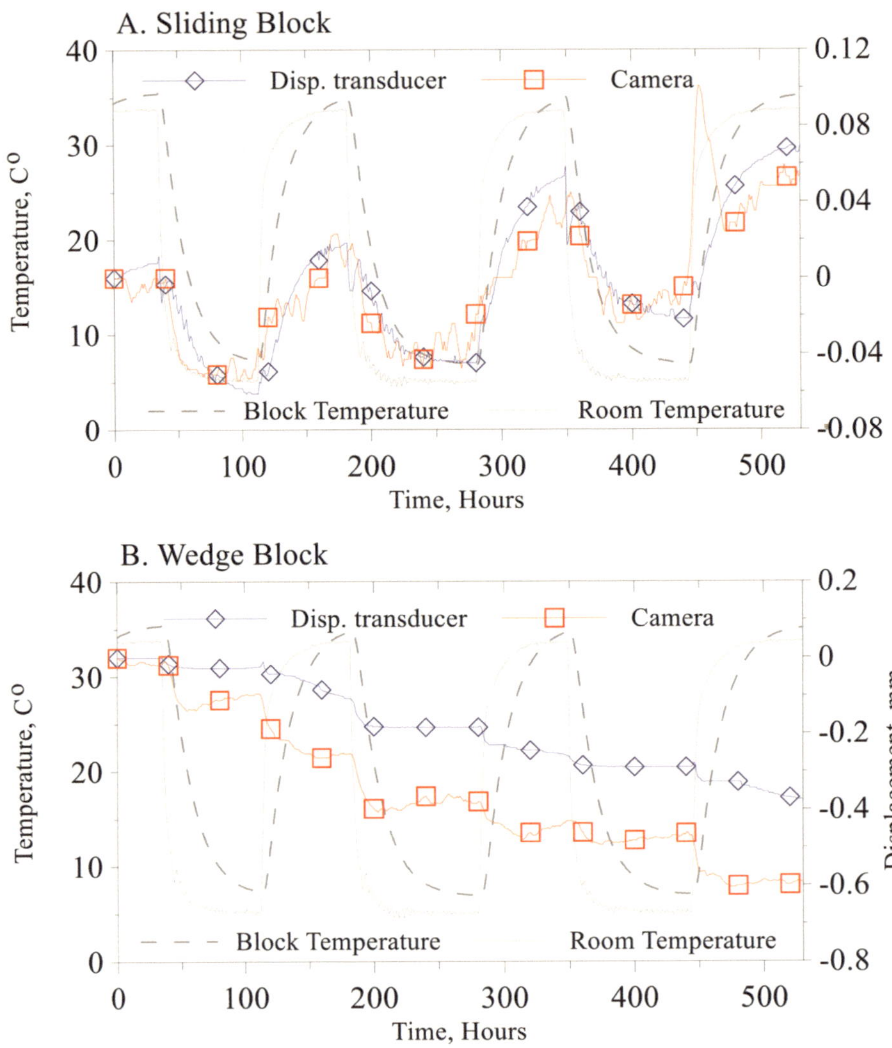


**Figure 8.**
*Physical model results. Displacement obtained by two different measurement methods (colored lines) is shown on the right y-axis, room and block temperature (gray lines) are shown on the left y-axis, for A) sliding block and B) wedge block.*

controlling parameters. We found that the numerical model is most sensitive to the chosen values of the normal stiffness in the contact between the blocks, and to the thermal expansion coefficient. The value of the thermal expansion coefficient that we used in the analysis represents linear expansion, while the problem in reality is affected by volumetric expansion. We assume this is one of the reasons for the differences between the laboratory measurements and the numerical results.

### 3.2 Field case study

After confirming that the numerical analysis captured the thermally induced ratchet mechanism, we used the code to test a case study in the field, returning to the case of Block 1 on the eastern slope of Masada. We used 3DEC to model the geometry

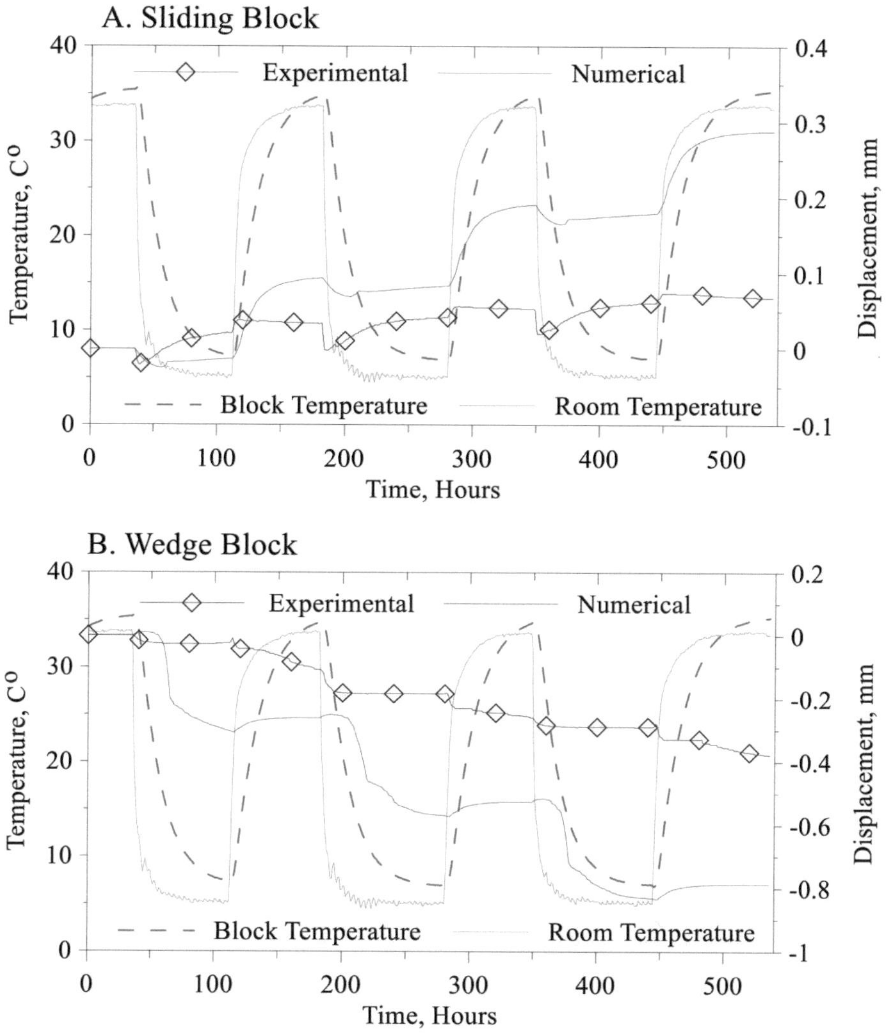


**Figure 9.**
*Comparison between numerical 3DEC results (purple lines) and physical model results obtained by the displacement transducer (blue lines). Displacement (colored lines) is shown on the right y-axis, and room and block temperature (gray lines) are shown on the left y-axis for A) sliding block and B) wedge block.*

of Block 1. The block rests on an inclined bedding plane (j1) with a dip angle of 20° toward the east (the free space). The maximum (peak) friction angle measured in the laboratory is 41°, and the residual (saw-cut) friction angle is 28° [1]. The block is separated from the mountain by two tensile joints—one joint, referred to as j2, is open, and fragments of rock can be seen inside the joint (**Figure 10A**), while the second joint, referred to as j3, is relatively closed (**Figure 10B**). Therefore, in the numerical model, we added a wedge block only inside j2 (**Figure 10D**). A temperature function, representing the air temperature as recorded at a nearby meteorological station, was applied to the free surfaces in the model. The numerical parameters used for the analysis are detailed in **Table 2**, assigned to the rock mass in Masada [1, 2]. A video describing the simulation over three seasonal cycles is attached to this chapter. The response of the block for three annual cycles is shown in **Figure 11**. The cumulative response of the block after three years is about 0.6 mm when considering a

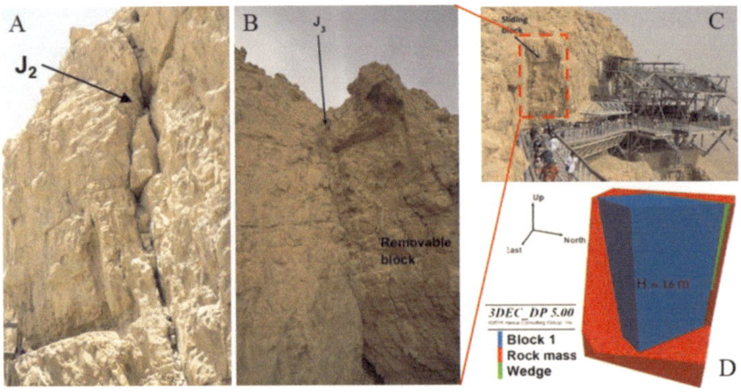


**Figure 10.**
*The analyzed Block 1 in Masada. A) J2 with the wedge fragments. B) The trace of J3 with no wedge blocks, C) the block near the cable car station, and D) the 3DEC model of the block.*

friction angle of the sliding surface of 41° and about 1.0 mm when considering a friction angle of 28°.

## 4. Discussion

### 4.1 The role of the wedge

Temperature fluctuations under conditions of biased forces are capable of causing geological materials to creep in accumulative plastic displacement [16]. Hence, it is assumed that rock blocks tend to creep on an inclined plane even without the wedge inside the joint, as also inferred by Gunzburgeret et al. [6]. To study the role of the wedge, we performed both the experiment and the numerical 3DEC analysis without a wedge block. The comparison between the two configurations, with and without a wedge, is shown in **Figure 12**. In our tested geometry, we infer that the wedge accelerates the cumulative slip rates approximately three times, compared to the situation where the tensile crack is empty for both the experiment and the numerical analysis.

We also analyzed Block 1 in Masada without a wedge block inside the tensile crack. Similar to the comparison made in the laboratory geometry, we numerically examined the effect of the wedge on the displacement rate of Block 1 in Masada. The results showed that the slip rate wedge acceleration was two times faster.

We used also the analytical solution to evaluate the annual displacement rate, as a function of the ratio between wedge and block length, for varied values of slope inclination (**Figure 13**). We use relevant physical and mechanical properties of Masada dolomite (**Table 2**). It seems that the mechanism becomes most effective when the length of the wedge is about a third of the length of the sliding block.

### 4.2 Daily vs. seasonal cycles

The rock mass response to temperature fluctuations occurs at relatively shallow depths because the temperature changes in the atmosphere do not penetrate too deeply into the rock. The extent of thermal penetration depends on the thermal conductivity of the material and the duration of the thermal cycle. Daily changes are

| | Parameter | Symbol | Units | Value | Source |
|---|---|---|---|---|---|
| Block geometry | Bedding plane | $J_1$ | dip/dip direction | 20/124 | [1] |
| | Joint 2 | $J_2$ | - | 84/107 | |
| | Joint 3 | $J_3$ | - | 75/052 | |
| | Face surface 1 | $f_1$ | - | 84/060 | |
| | Face surface 2 | $f_2$ | - | 90/126 | |
| | Block volume | $V$ | $m^3$ | 563 | |
| Material properties | Elastic modulus | $E$ | GPa | 40 | [2] |
| | Poisson's ratio | $\nu$ | - | 0.2 | |
| | Bulk density | $\rho$ | $kg/m^3$ | 2600 | |
| | Block mass | $W$ | $10^6 kg$ | 1.465 | |
| Joints properties | Peak friction angle | $\phi_{peak}$ | ° | 41 | [1] |
| | Saw-cut friction angle | $\phi_{saw}$ | ° | 28 | |
| | Residual friction angle | $\phi_{res}$ | ° | 23 | |
| | Normal stiffness | $K_n$ | GPa/m | 5 | [2] |
| | Shear stiffness | $K_s$ | GPa/m | 1 | |
| Thermal properties | Thermal expansion coefficient | $\alpha$ | $10^{-6}$/°C | 6-8 | [2] |
| | Thermal conductivity | $\lambda$ | W/m/K | 1.7 | |
| | Thermal diffusivity | $D_T$ | $10^{-7} m^2/sec$ | 8.07 | calculated |
| | Specific heat capacity | $C_p$ | J/kg/K | 810 | [15] |

**Table 2.**
*Properties of the removable block in Masada.*

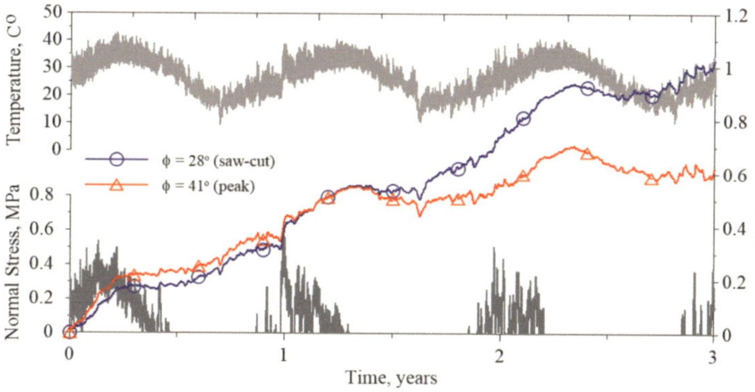


**Figure 11.**
*Thermally induced displacements in Block 1 in Masada, as computed with 3DEC for peak and residual (saw-cut) friction. The applied temperature to the block boundaries is shown on the upper panel; the normal compressive stresses that evolve at the back of the wedge in response to thermal oscillations are shown on the lower panel.*

sensed at depths of only a few cm, while seasonal changes can be discerned several meters into the rock. Therefore, we assumed that the larger the blocks on the slope, the more likely they will be affected by seasonal changes rather than daily changes. To test this on Block 1 at Masada, we implemented the temperature function in two ways:

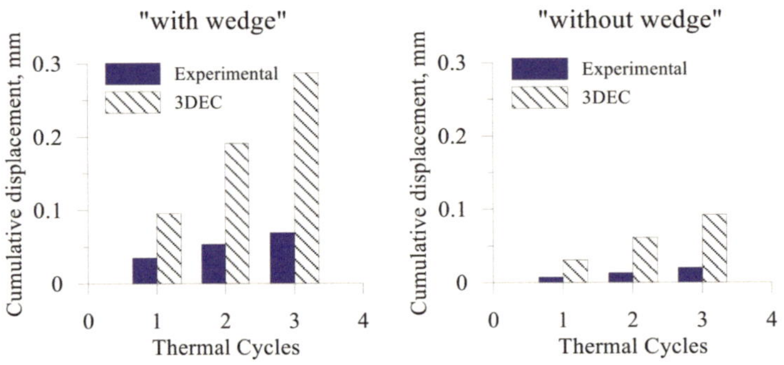


**Figure 12.**
*Cumulative displacements for each thermal cycle as obtained with the numerical and physical models for the "Ratcheting" (left) and crawling (right) failure mechanisms.*

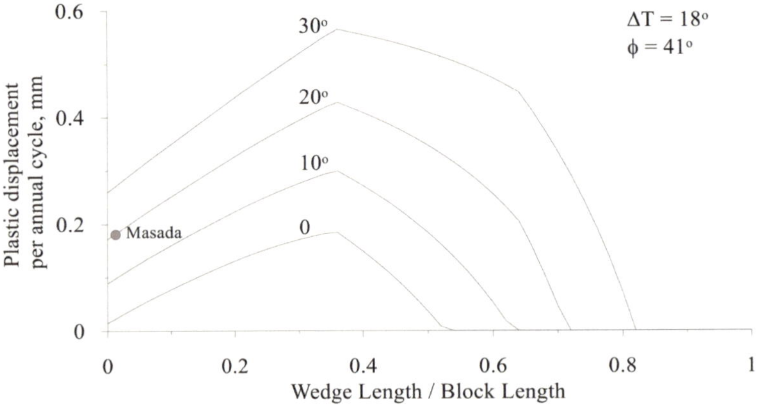


**Figure 13.**
*One cycle plastic displacement for several plane inclination angles for Masada dolomite, as calculated by the analytical expressions suggested by Pasten [9].*

1) using a full record that also contains the daily measurements, and 2) using a smoothed record that contains only the seasonal changes. The results are shown in **Figure 14**. It is indicated that the daily changes do not affect the cumulative response of the block; namely, the cycle duration is too short to affect the heating and cooling of the block. However, several consecutive days of extreme temperature can lead to a cumulative response, as seen in the gray areas in **Figure 14** and in **Figure 1D**.

### 4.3 Rock types affected by the mechanism

Thermally induced rock displacement depends mainly on the rock material's thermal conductivity and the thermal expansion coefficient. To examine which rock types are more responsive to the ratchet mechanism, we performed a numerical analysis using 3DEC on the system geometry of the blocks representing the physical model. We used a different type of rock in each analysis, with physical properties taken from the literature (**Table 3**). The results of the analysis for different types of rock are shown in **Figure 15**. It is clearly that sandstone reacts to the thermally induced ratchet mechanism more extensively, probably because of the high value of the thermal expansion coefficient.

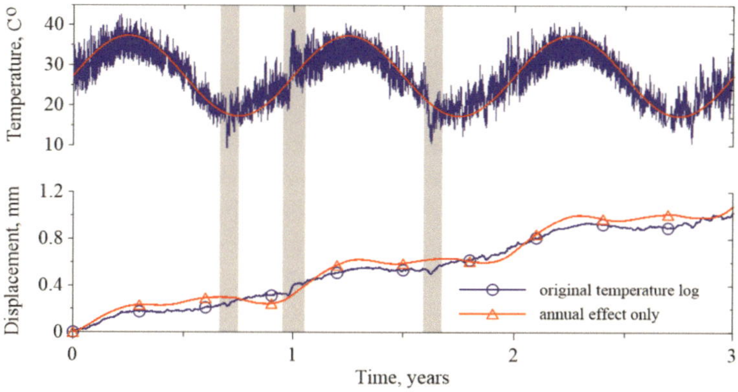


**Figure 14.**
*The influence of short-term thermal fluctuations on block displacement. Top) Temperature input for the simulations. Bottom) Block displacement as computed with 3DEC for the two input temperature records.*

| Parameter | Symbol | Units | Granite | Basalt | Sandstone | Dolomite |
|---|---|---|---|---|---|---|
| Elastic modulus | $E$ | GPa | 50 [17] | 70 [17] | 20 [18] | 48 [19] |
| Poisson's ratio | $\nu$ | - | 0.2 [20] | 0.28 [20] | 0.25 [21] | 0.26 [19] |
| Bulk density | $\rho$ | Kg/m$^3$ | 2650 [17] | 2870 [22] | 2640 [22] | 2550 [19] |
| Thermal expansion coefficient | $\alpha$ | $10^{-6}$/°C | 8 [23] | 5 [24] | 11[25] | 8 [23] |
| Thermal conductivity | $k$ | W/m/K | 2.5 [26] | 1.2 [27] | 3.3 [24] | 4.5 [24] |
| Specific heat capacity | $C_p$ | J/kg/K | 840 [26] | 840 [28] | 900 [28] | 900 [28] |
| Thermal diffusivity (calculated) | $\kappa$ | $10^{-6}$m$^2$/sec | 1.12 | 0.50 | 1.39 | 1.96 |

**Table 3.**
*Typical thermomechanical properties of some rocks.*

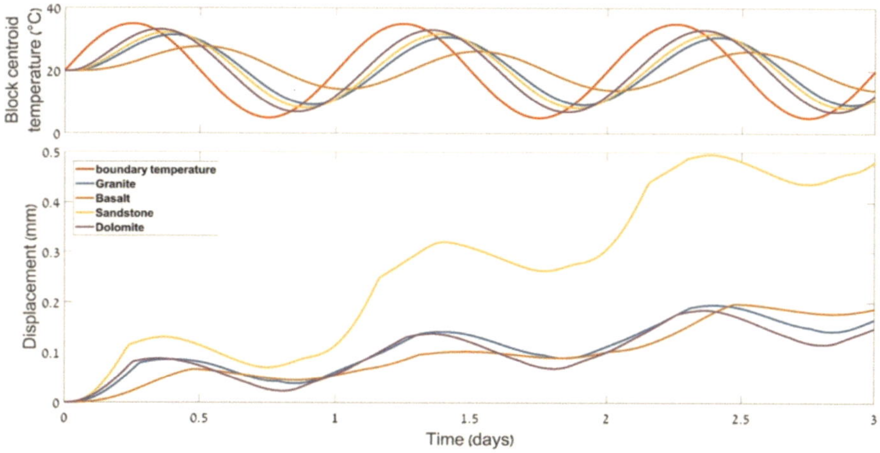


**Figure 15.**
*Response to the thermal ratcheting mechanism; a comparison between different rock types. The analyses were done using 3DEC on the geometry and thermal boundary conditions used on the physical model inside the climate-controlled room.*

## 5. Summary

This chapter presented observations showing that a jointed rock mass responds to temperature fluctuations. First, we presented a few examples of measurements observed at several monitoring stations around the world. The difficulty in field measurements is to isolate the effect of temperature from other environmental factors, such as water pressure in joints due to a freezing and thawing process. For this reason, Mount Masada is a preferred site for testing thermal mechanisms because of the prevailing desert conditions, where there is usually no pore pressure or freezing of underground water.

Then, a ratchet mechanism was used to explain how temperature changes can lead to cumulative displacement along a rock slope. Laboratory results of models of different sizes were presented, and the effect of the wedge size was demonstrated. Numerical analyses using 3DEC illustrate the effect of the wedge inside the tensile crack. We have seen that the wedge generally accelerates the cumulative response of the block system, and the maximum response occurs where the wedge length in about a third of the length of the sliding block.

The numerical analyses also make it possible to examine the contribution of daily and annual fluctuations. We saw that in the model of the block in Masada, the daily temperature fluctuations do not affect the block due to its large size. However, a sequence of several days with extreme temperatures for the season can lead to a cumulative response. Finally, based on the numerical analysis, rocks made of material with a high thermal expansion coefficient will respond more to the thermal mechanism.

## Acknowledgements

I thank Prof. Yossi Hazor for his support throughout the research. The laboratory work and the numerical analyses were carried out by Aviran Feldheim and Yuval Keissar, with the support of the Israel Science Foundation under contract no. ISF 1442/13. Christine Detournay from Itasca Consulting Group is thanked for her contribution.

## Author details

Dagan Bakun-Mazor
Shamoon College of Engineering, Beer-Sheva, Israel

*Address all correspondence to: daganba@sce.ac.il

## References

[1] Hatzor YH. Keyblock stability in seismically active rock slopes – Snake Path cliff, Masada. Journal of Geotechnical and Geoenvironmental Engineering. 2003;**129**(8):697-710

[2] Bakun-Mazor D, Hatzor YH, Glaser SD, Santamarina JC. Thermally vs. seismically induced block displacements in Masada rock slopes. International Journal of Rock Mechanics and Mining Sciences. 2013;**61**:196-211

[3] Vicko J, Greif V, Grof V, Jezny M, Petro L, Brcek M. Rock displacement and thermal expansion study at historic heritage sites in Slovakia. Environmental Geology. 2009;**58**(8):1727-1740

[4] Gischig VS, Moore JR, Evans KF, Amann F, Loew S. Thermomechanical forcing of deep rock slope deformation: 2.The Randa rock slope instability. Journal of Geophysical Research-Earth Surface. 2011;**116**:F04011

[5] Mufundirwa A, Fujii Y, Kodama N, Kodama J. Analysis of natural rock slope deformations under temperature variation: A case from a cool temperate region in Japan. Cold Regions Science and Technology. 2011;**65**(3): 488-500

[6] Gunzburger Y, Merrien-Soukatchoff V, Guglielmi Y. Influence of daily surface temperature fluctuations on rock slope stability: Case study of the Rochers de Valabres slope (France). International Journal of Rock Mechanics and Mining Sciences. 2005;**42**(3):331-349

[7] Fiorucci M, Martino S, Bozzano F, Prestininzi A. Comparison of approaches for data analysis of multi-parametric monitoring systems: Insights from the Acuto test-site (Central Italy). Applied Sciences. 2020;**10**(21):7658

[8] Taboada A, Ginouvez H, Renouf M, Azemard P, editors. Landsliding generated by thermomechanical interactions between rock columns and wedging blocks: Study case from the Larzac Plateau (Southern France). EPJ Web of Conferences (EDP Sciences). 2017

[9] Pasten C. Geomaterials Subjected to Repetitive Loading: Implications on Energy Systems. Atlanta, GA: Georgia Institute of Technology; 2013

[10] Bakun-Mazor D, Keissar Y, Feldheim A, Detournay C, Hatzor YH. Thermally-induced Wedging–Ratcheting failure mechanism in rock slopes. Rock Mechanics and Rock Engineering. 2020;**53**(6):2521-2538

[11] Pasten C, García M, Cortes D. Physical and numerical modelling of the thermally induced wedging mechanism. Geotechnique Letters. 2015;**5**(3):186-190

[12] Greif V, Simkova I, Vlcko J. Physical model of the mechanism for thermal wedging failure in rocks. In: Sassa K, Canuti P, Yin Y, editors. Landslide Science for a Safer Geoenvironment. Cham: Springer; 2014. pp. 45-50

[13] Cundall PA. Formulation of a 3-dimensional distinct element model.1. A scheme to detect and represent contacts in a system composed of many polyhedral blocks. International Journal of Rock Mechanics and Mining Science and Geomechanics Abstracts. 1988; **25**(3):107-116

[14] Hart R, Cundall PA, Lemos J. Formulation of a 3-dimensional distinct element model .2. Mechanical calculations for motion and interaction of a system composed of many polyhedral blocks. International Journal of Rock Mechanics and Mining Science

and Geomechanics Abstracts. 1988; **25**(3):117-125

[15] Rohsenow WM, Hartnett JP, Cho YI. Handbook of Heat Transfer. New York: McGraw-Hill; 1998

[16] Pastén C, Castillo E, Chong S-H. Thermo-mechanical ratcheting in soil–structure interfaces. Acta Geotechnica. 2019;**14**(5):1561-1569

[17] Turcotte DL, Schubert G. Geodynamics. New York, NY: Cambridge University Press; 2014

[18] Vasarhelyi B. Some observations regarding the strength and deformability of sandstones in dry and saturated conditions. Bulletin of Engineering Geology and the Environment. 2003; **62**(3):245-249

[19] Hatzor Y, Palchik V. The influence of grain size and porosity on crack initiation stress and critical flaw length in dolomites. International Journal of Rock Mechanics and Mining Sciences. 1997;**34**(5):805-816

[20] Vutukuri V, Lama R, Saluja S. Handbook on mechanical properties of rocks. Testing techniques and results. International Journal of Rock Mechanics and Mining Sciences. 1974; **11**(11):A218

[21] Gercek H. Poisson's ratio values for rocks. International Journal of Rock Mechanics and Mining Sciences. 2007; **44**(1):1-13

[22] Jaeger JC, Cook NG, Zimmerman R. Fundamentals of Rock Mechanics. 4th ed. Oxford, UK: Blackwell Publishing Ltd; 2007

[23] Franklin J, Dusseault M. Rock Engineering Applications. New York: McGraw-Hill; 1991. p. 600

[24] Naeser ND, Naeser CW, McCulloh TH. The application of fission-track dating to the depositional and thermal history of rocks in sedimentary basins. In: Thermal History of Sedimentary Basins. New York, NY: Springer; 1989. pp. 157-180

[25] Engineering-ToolBox. Thermal Expansion - Linear Expansion Coefficients: online; 2003. Available from: https://www.engineeringtoolbox.com/linear-expansion-coefficients-d_95.html

[26] Heuze FE. High-temperature mechanical, physical and thermal properties of granitic rocks—a review. International Journal of Rock Mechanics and Mining Science and Geomechanics Abstracts. 1983;**20**(1):3-10

[27] Blesch C, Kulacki F, Christensen R. Application of integral methods to prediction of heat transfer from a nuclear waste repository. Columbus (USA): Ohio State Univ; 1983

[28] Eppelbaum L, Kutasov I, Pilchin A. Thermal properties of rocks and density of fluids. In: Applied Geothermics. Berlin, Heidelberg: Springer; 2014. pp. 99-149